WOW
NUMEROLOGY
OR, WHAT
PYTHAGORAS
WROUGHT

NUMEROLOGY

OR, WHAT PYTHAGORAS WROUGHT

UNDERWOOD DUDLEY

Mathematical Association of America

Library of Congress Catalog Card Number 97-74345

ISBN 0-88385-524-0

Printed in the United States of America

Current Printing (last digit):
10 9 8 7 6 5 4 3 2 1

SPECTRUM SERIES

Published by
THE MATHEMATICAL ASSOCIATION OF AMERICA

SPECTRUM SERIES

The Spectrum Series of the Mathematical Association of America was so named to reflect its purpose: to publish a broad range of books including biographies, accessible expositions of old or new mathematical ideas, reprints and revisions of excellent out-of-print books, popular works, and other monographs of high interest that will appeal to a broad range of readers, including students and teachers of mathematics, mathematical amateurs, and researchers.

All the Math That's Fit to Print, by Keith Devlin
Circles: A Mathematical View, by Dan Pedoe
Complex Numbers and Geometry, by Liang-shin Hahn
Cryptology, by Albrecht Beutelspacher
Five Hundred Mathematical Challenges, Edward J. Barbeau, Murray S. Klamkin, and William O. J. Moser
From Zero to Infinity, by Constance Reid
I Want to be a Mathematician, by Paul R. Halmos
Journey into Geometries, by Marta Sved
JULIA: a life in mathematics, by Constance Reid
The Last Problem, by E. T. Bell (revised and updated by Underwood Dudley)
The Lighter Side of Mathematics: Proceedings of the Eugène Strens Memorial Conference on Recreational Mathematics & its History, edited by Richard K. Guy and Robert E. Woodrow
Lure of the Integers, by Joe Roberts
Magic Tricks, Card Shuffling, and Dynamic Computer Memories: The Mathematics of the Perfect Shuffle, by S. Brent Morris
Mathematical Carnival, by Martin Gardner
Mathematical Circus, by Martin Gardner
Mathematical Cranks, by Underwood Dudley
Mathematical Magic Show, by Martin Gardner
Mathematics: Queen and Servant of Science, by E. T. Bell
Memorabilia Mathematica, by Robert Edouard Moritz
New Mathematical Diversions, by Martin Gardner
Numerical Methods that Work, by Forman Acton
Numerology or What Pythagoras Wrought, by Underwood Dudley
Out of the Mouths of Mathematicians, by Rosemary Schmalz
Penrose Tiles to Trapdoor Ciphers... and the Return of Dr. Matrix, by Martin Gardner
Polyominoes, by George Martin
The Search for E. T. Bell, also known as John Taine, by Constance Reid
Shaping Space, edited by Marjorie Senechal and George Fleck
Student Research Projects in Calculus, by Marcus Cohen, Edward D. Gaughan, Arthur Knoebel, Douglas S. Kurtz, and David Pengelley
The Trisectors, by Underwood Dudley
The Words of Mathematics, by Steven Schwartzman

MAA Service Center
P. O. Box 91112
Washington, DC 20090-1112
800-331-1MAA FAX 301-206-9789

Contents

CHAPTER 1
Introduction

This book is about numbers. Not about numbers in their workaday role as counters (send $3.50 plus $4.95 for postage and handling, a total of $8.45), or as mathematical objects (845 is a sum of two squares in three different ways: $29^2 + 2^2$, $26^2 + 13^2$, and $22^2 + 19^2$), but as things about which can be said,

> No, I wouldn't go so far as some of my fellow calculators and indiscriminately welcome all numbers with open arms: not the heavy-handed rough-and-tough bully 8 or the sinister 64 or the arrogant, smug self-satisfied 36. But I do admit to a very personal affection for the ingenious, adventurous 26, the magic, versatile 7, the helpful 37, the fatherly, reliable (if somewhat stodgy) 76... [3, pp. xii–xiii]

and

> 9 is a wonderful being of whom I felt almost afraid, 8 I took for his wife, and there used always to seem a fitness in 9×9 being so much more than 8×8. 7 again is masculine; 6, of no particular sex but gentle and straightforward; 3 a feeble edition of 9, and generally mean; 2 young and sprightly; 1 a common-place drudge. [2, p. 253]

For some people, numbers do much more than merely count and measure. For some people, numbers have *meanings*, they have *inwardnesses*, they can be magic and versatile, or young and sprightly. I am not one of those people, since I think that numbers have quite enough to do as it is, but for the crowd of number mystics, numerologists, pyramidologists, number-of-the-beasters, and others whose ideas and work will be described in the following chapters, numbers have powers far out of the ordinary.

Number mysticism got its start in ancient Greece, with Pythagoras and the Pythagoreans in the sixth century B.C. Before that, numbers were just numbers, things to count with. The Pythagoreans made some discoveries about numbers—for instance, that the sum of odd numbers starting at 1 is always a square,

$$1 + 3 = 2^2, \quad 1 + 3 + 5 = 3^2, \quad 1 + 3 + 5 + 7 = 4^2, \ldots$$

which so impressed them that they reportedly came to the mystical conclusion that "all is number." If all is number, then numbers are worth investigating, and that is what the Pythagoreans did to the limit of their abilities, both mathematically and mystically. Both mathematics and number mysticism have been marching along ever since, though no longer together—they parted company forever within one hundred years of the death of Pythagoras.

Mysticism is a nonrational method of getting at truth. Ours is a rational era (even if it seems to become less so every year), and we can lose sight of the existence of truths that are not arrived at by reason: emotional truths, spiritual truths, even physical truths. Some truths cannot be described in words, nor arrived at by reason. Love provides one example. For another, can you describe a sneeze in words? Or what it feels like? If you can, you are a better wordsmith than I. Some truths must be *felt*.

Number mystics, by feeling properties of numbers, gain mystical insights into the nature of the universe. Not being a number mystic myself, I cannot describe them, but I could not describe them no matter how gifted a mystic I was since, by definition, mystical experiences are ineffable. Those who have them are fortunate. There is nothing wrong with mysticism.

On the other hand, everything is wrong with numerology. Numerologists purport to *apply* number mysticism. That is, they take mystical properties of numbers—2 is cold, say, and wet—and attach them to things and people. If your number is 2 (numbers can easily be assigned to people, in many different ways), then *you* are cold and wet, whether you know it or not.

This is standing mysticism on its head. For a number mystic, numbers are tools, means of gaining understanding. For a numerologist, numbers are the masters, dictating the nature of the world. Numerologists assert that numbers tell you where it would be best to live, who you should marry, even at what time you should arrive for an appointment. Numbers do not do this. It is not their job. Numbers have power, but not that kind of power.

This is a thread that runs through numerology, pyramidology, and the many other misuses of numbers that are described in this book. What they have in common is the belief that things happen *because numbers make them happen.* The pyramidologist measures his pyramid and says that the world will end on August 20, 1953 *because* of his measurements (chapter 25). A neo-Pythagorean says that Greeks carved the Easter Island statues *because* his numbers tell him so (chapter 4). A distinguished Oxford University scholar asserts that one of Shakespeare's sonnets is irregular *because* 28 is a triangular number (chapter

17). The stock market behaves as it does *because* Fibonacci numbers make it go up and down (chapter 33).

Another thread is the numerologists' refusal to believe that patterns can occur by accident. Human beings are very good at seeing patterns, and sometimes they see patterns that no one made but exist only by chance. The Bible is full of 7s, one author tells us, while another finds 13s. Yet another finds squares, and a fourth finds triangular numbers. Each of the four says that the numbers are there because God put them there. It is possible that God takes delight in confounding poor, limited humans with obscure puzzles, and it is possible that the Bible is full of 17s or 23s that were put there on purpose and that no one has noticed yet, but I doubt it. The 7s, 13s, squares, and triangular numbers are all there by chance. Like numbers, chance has power.

Others find that people they dislike bear the number of the beast, 666. They are thus bad, *because* of the number. Or, they think that they find 666 in the bar code that is on almost everything that we buy and, *because* of the number, deduce the existence of a vast conspiracy. Biorhythmists say that we are all oscillating in cycles of lengths 23, 28, and 33 days and act as we do *because* of the numbers. There is something about numbers that can turn the head.

What follows in this book is a description of these and other manifestations of number mysticism and numerology. The lessons to be learned are that

numbers have power

but

numbers do not control events

and

coincidences happen.

These facts are so obvious that they hardly need stating; so the question is, why read any further? The answer is that by so doing you can learn about something that you may not be aware of, the world of number mysticism and those infected by it: the pyramid-measurers, bible-numberists, Elliott Wavers, and so on. It is, I think, a colorful and interesting world, and worth knowing a little about.

There is, by the way, no other book devoted to this subject. *Numerology* by E. T. Bell [1] contains nothing on modern numerology, and other books with that word in their titles tend to be written by numerologists. Since this book covers more than numerology, a more descriptive title for it would be its subtitle, *What Pythagoras Wrought*, but that would be less informative. Also, it is my hope that copies of it will turn up on the New Age shelves of used book stores, where they may fall into the hands of those expecting something different. The shock may do them good.

I wish to thank those who provided me with material or other help, in particular Arthur Benjamin, James Bidwell, I. J. Good, Richard Guy, Michael Keith, David Singmaster, Diane Spitler, Ian Stewart, Michael Stueben, and especially Martin Gardner, who allowed me to inspect his wonderful files. I am also indebted to the Fisher Fund of DePauw University, which provided a semester free of teaching duties.

References

1. Bell, E. T., *Numerology*, Century, New York, 1933, reprinted by Hyperion Press, Westport, Connecticut, 1979.
2. Galton, Francis, Visualised numerals, *Nature* **21** (1879–1880), 252–256.
3. Smith, Stephen B., *The Great Mental Calculators*, Columbia University Press, New York, 1983.

CHAPTER 2

Pythagoras

Who was Pythagoras? What did he do? Why did he do it? Hard questions all, with no easy answers.

The person in the street, if aware of Pythagoras at all, would probably say that he proved the Pythagorean theorem (which has, more or less, become part of general knowledge) and then not be able to think of anything else. Those more acquainted with mathematics might be able to remember that he discovered that $\sqrt{2}$ is an irrational number and sacrificed an ox (or a hecatomb—100—thereof) in celebration, and a few might bring up his discovery of musical harmonies. Beyond that, I think that most people, even the mathematically inclined, would draw blanks.

This is too bad, because Pythagoras is one of the most important people in human history. That is, if you measure importance by the magnitude of the effect that a person has on people who come after, which seems to be as good a measure as any. What Pythagoras did was to turn number mysticism loose on the world, along with its corollaries—numerology, pyramidology, and many others. This was no small achievement. Number mysticism, from belief in the unluckiness of 13 on up, has been spread far and wide, and is still spreading. There is no way to determine the total number of hours that the human race has devoted to number mysticism and its offspring, but I am sure it is a significant fraction of the number of hours the race has devoted to mathematics. The fraction may even be greater than one.

Pythagoras was responsible for number mysticism. The writings of the civilizations that came before the ancient Greeks, those of the Egyptians and Babylonians, contain no number mysticism. Numbers for them were numbers, useful and necessary for numbering and calculation, but they did not have inner meanings, souls, vibrations, or any of the mystical paraphernalia that have trailed along with them since the time of Pythagoras.

It is possible, even likely, that Pythagoras absorbed number-mystical ideas from the East (a traditional hotbed of mysticism), from sources now lost; but if he had not done what he did, they might have stayed lost. It is also possible that number mysticism would have emerged sooner or later even if Pythagoras had never lived—I am sure, though partly for mystical reasons, that it was in fact inevitable—nevertheless, he was responsible, and so he gets the credit, or blame. Someone else might have hit on relativity if Einstein had not, but we do not honor Einstein any less for that. Wherever number mysticism came from, or whether someone else would have spread it throughout the world, it is still the case that Pythagoras did it.

Further, I assert, it was *all* that he did. He did not prove the Pythagorean theorem, he did not discover that $\sqrt{2}$ was irrational, and I doubt that he ever did any research into musical harmonies. I maintain that Pythagoras was a successful cult leader, and that is essentially all that he was. I think he had a modern parallel in Senator Joseph R. McCarthy, who was successful in raising a large and very powerful anticommunist cult in the 1950s. McCarthy's cult was not as successful as Pythagoras's, partly because McCarthy was, among other unpleasant things, a drunk. McCarthy selected anticommunism as his issue not because of any intense intellectual or emotional attachment to it, but because he needed an issue and anticommunism was handy, not in use, and looked promising. I think that Pythagoras acted similarly.

I am not the only one to think so. Heraclitus (c. 540–480 B.C.), who was around at the time, called Pythagoras "the chief of swindlers" and said that he made wisdom of his own by picking things out of the writings of others. Carl Huffman, from whose article that is taken, says,

> the overall tone of Heraclitus' reports... is clear. He regards Pythagoras as a charlatan of some sort. [3]

A charlatan or not (and I think he was not—sincerity is a characteristic of successful cult leaders), his cult attracted people of intelligence and of a mathematical as well as mystical turn of mind, and it is to them that the mathematical accomplishments attributed to Pythagoras are due. It is natural for members of a cult to want to glorify their Master.

Let us go to an encyclopedia to see what it says about Pythagoras:

> Pythagoras of Samos, c. 560–c. 480 BC, was a Greek philosopher and religious leader who was responsible for important developments in the history of mathematics, astronomy, and the theory of music. He migrated to Croton and founded a philosophical and religious school there that attracted many followers. Because no reliable contemporary records survive, and because the school practiced both secrecy and communalism, the contribution of Pythagoras and those of his followers cannot be distinguished. Pythagoreans

> believed that all relations could be reduced to number relations ("all things are numbers"). This generalization stemmed from certain observations in music, mathematics, and astronomy.
>
> The Pythagoreans noticed that vibrating strings produce harmonious tones when the ratios of the lengths of the strings are whole numbers, and that these ratios could be extended to other instruments. They knew, as did the Egyptians before them, that any triangle whose sides were in the ratio 3:4:5 was a right-angled triangle. The so-called Pythagorean theorem, that the square of the hypotenuse of a right triangle is equal to the sum of the squares of the other two sides, may have been known in Babylonia, where Pythagoras traveled in his youth; the Pythagoreans, however, are usually credited with the first proof of this theorem.
>
> In astronomy, the Pythagoreans were well aware of the periodic numerical relations of the heavenly bodies. The CELESTIAL SPHERES of the planets were thought to produce a harmony called the music of the spheres. Pythagoreans believed that the earth itself was in motion. The most important discovery of this school—which upset Greek mathematics, as well as the Pythagoreans' own belief that whole numbers and their ratios could account for geometrical properties—was the incommensurability of the diagonal of a square with its side. This result showed the existence of IRRATIONAL NUMBERS.
>
> Whereas much of the Pythagorean doctrine that has survived consists of numerology and number mysticism, the influence of the idea that the world can be understood through mathematics was extremely important to the development of science and mathematics. [2]

Except for the story about Pythagoras traveling in Babylonia (for which there is no evidence other than tradition) and information that the ancient Egyptians knew about the 3-4-5 right triangle (they didn't), this account is accurate. Not much can be said about Pythagoras because hardly anything is known about him for certain. He was from Samos, he and his followers lived in Croton in southern Italy, he and they were forced out, he went to Metapontum, where he died. That is certain.

There is quite a bit that is not certain. As with other cult leaders, many stories are told about Pythagoras and his teachings. For example:

- Pythagoras appeared in both Croton and Metapontum at the same hour on the same day.
- When he crossed a river, it gave him the greeting, "Hail Pythagoras."
- He bit and killed a poisonous snake.
- He taught that you should not eat the heart of an animal, or beans.
- He ate no meat.
- New members of the Pythagorean society were forbidden to speak for five years.

- He said that you should not wear wool in a temple.
- When a man was beating a dog he said "Stop, don't keep beating it, it is the soul of a friend of mine, I recognize his voice."

Many of these are absurd, the others are not reliable, and there is no reason to suppose that other assertions about his life and works are any more dependable. The reason is that

> most of the sources concerning Pythagoras' life, activities, and doctrines date from the third and fourth centuries A.D., while the few more nearly contemporary (fourth and fifth centuries B.C.) records of him are often contradictory, due in large part to the split that developed among his followers after his death. [4, vol. 4, pp. 2074–75]

(The split was between the mathematical and mystical parties of Pythagoreans.)

We can be sure, though, that he had a significant number of followers; he led a cult, with rules and doctrines, and members devoted to its leader. It is not clear whether original Pythagoreanism, led by Pythagoras, had any mathematical or numerical component. It may have been that his personality was so strong that those who were susceptible to the joys of cult membership—the abdication of responsibility, the feeling of being in a special group, the chance to be near the semi-divine leader—would have followed him whatever his doctrines were. It is quite possible that original Pythagoreanism had no doctrines at all besides adoration of Pythagoras, except for that of the transmigration of souls, for which there is contemporary evidence. He may even have eaten beans. In [4] we find

> this rule [about beans] was never laid down by Pythagoras, but was rather an interpolation by some medieval busybody, and in fact there was nothing Pythagoras liked better than a pot of Great Northern beans simmered with a bit of ham hock. [p. 18]

The novel from which that is taken is as reliable a source as those in which the other stories about Pythagoras appeared.

Almost by definition, cults are not democratic organizations, and Pythagoreanism was no exception. Tradition has it that there were two classes of members, the *akousmatikoi*, or listeners, and the *mathematikoi*, who were more advanced. (Since the Greek root *mathe-* means "learning," the name of the advanced group says nothing about their knowledge of numbers or mathematics.) Unfortunately for the Pythagoreans, the tide of democracy was rising in Greece, and this contributed to the persecution of the society. Pythagorean societies probably did not survive the death of Pythagoras for much more than fifty years.

E. T. Bell, who never let facts, or the lack of them, get in the way of a good story, wrote a description (entirely imaginary, though Professor Bell did not say so) of the Pythagorean life:

> A few details will suffice to indicate the kind of life the Pythagoreans lived and the rigors of the discipline to which they submitted. The harshness of a listener's probationary period was extreme. For three inhospitable years the would-be mathematician was hazed unmercifully. Should he venture an opinion or offer a harmless remark, his seniors first rudely contradicted him, then smothered him in ridicule and contempt. If the candidate was worthy, a year of such browbeating was usually enough to inculcate the virtues of silence and forbearance.
>
> A meager diet, with no animal food except a scrap now and then left over from sacrifices to the insatiable gods, enforced the lesson of moderation. The generous wine which cheered the common man was prohibited, except for a sip or two before going to bed purely as health insurance. Any tendency towards gourmandizing was checked by seating the patient comfortably at a loaded banquet table, letting him savor the appetizing aromas in anticipatory ecstasy till he reached for his favorite dish, when it was snatched away. His garments were scanty and coarse, but sufficient to keep him rugged. Even the solace of oblivion was denied him until he learned to get along with three or four hours' sleep and like it. Any little comforts he might have brought with him to soften his purgatory followed all his more substantial possessions into the common stock, and he enjoyed them no more.... When a candidate finally got used to the life he found it not much harsher than the basic training in a barbarous military camp.
>
> While the body was being toughened the mind was by no means neglected. Long before sunrise the day began with semi-religious exercises. Lofty metaphysical poetry and elevating mathematical music hardened the auditors for a solitary walk of meditation before their cheerless breakfast. During this walk each planned his day. Good intentions were balanced against performance at sunset. Should some unhappy wretch do some things which he ought not to have done, or leave undone some things which he ought to have done, he penalized himself appropriately the next day.
>
> The morning bread and water was followed by a short period of relaxation to prepare for the real rigors of the day. All gathered for a friendly chat. The few who had earned the privilege of expressing their minds spoke softly and sparingly while the other listened and said nothing at all. This unilateral style of conversation was designed to further the capital ideal of producing submissive minds in disciplined bodies.
>
> The Pythagoreans were among the earliest discoverers of the physiological fact that hard physical work is a slow poison destructive of creative thinking. Being relieved by their slaves of the necessity of indulging in brutalizing labor, they kept themselves fit by judicious doses of cultural athletics. Wrestling bouts, running, javelin tossing, and similar sports sharpened their appetites for the tasteless evening meal of bread, honey, and water. The math-

> ematicians, supposed to be above such frailties of the flesh, got only pure cold water and not too much of that.
>
> Any mathematicians still awake after their unexciting repast—consumed in silence—turned to the administration of the Brotherhood's domestic and foreign affairs. The survivors of this tedious ordeal refreshed themselves with protracted religious exercises of mystical solemnity, took a cold bath, and fell upon their stony beds. Up again some hours before daybreak, they plunged once more into the endless round of music, meditation, talking or listening, solitary promenades, introspection, unappetizing meals, numerology, science, mathematics, religion, athletics, metaphysics, bathing, and just enough sleep to prevent them from dozing off on their feet. It was no life for a sybarite. [1, pp. 117–19]

A good story, better than any I could have made up. It probably has a good deal of truth in it, though I suspect that the Pythagorean life was not entirely as harsh as Professor Bell painted it. Cults need to attract members, not repel them.

Professor Bell also had a portrait of Pythagoras, woven out of cloth as whole as that which provided the previous excerpt:

> At the peak of the Brotherhood's prosperity some two hundred families (other estimates give three times as many) lived together more or less harmoniously under the fatherly supervision of Pythagoras. As for the master himself, he enjoyed every moment of his undisputed authority. Numbers were not the only mysteries he understood better than any of his disciples. In cultist psychology he is still without an equal in all the long and varied history of cults. Always aloof, even when conferring with his brother mathematicians, he seldom spoke unless he had something mystifying to communicate. Taciturnity seems to have been a passion with him, if not for himself, then certainly for his followers. To ensure a properly respectful acceptance for his teachings, he imposed a silence of from three to five years on listeners newly promoted to the grade of mathematician. His disciples seldom saw him, but when they did they were overwhelmed by the majesty of his bearing. Like the master of showmanship he was, Pythagoras always chose the unexpected moment to exhibit himself. His rare appearances were rendered sufficiently godlike and remote by a voluminous white robe, a crown of golden leaves, and his full white beard. To heighten the mystery of his more recondite doctrines he intoned his most confidential utterances behind a curtain. The organ voice, accompanied by melodious chords struck out with bold abandon on his lyre, convinced the more credulous of his auditors that they were hearing Apollo. Pythagoras never made the mistake of stepping from behind the curtain when the last note of his musical discourse had perished in the quivering silence.
>
> When the curtain began to wear threadbare the master retired with his lyre to the Grotto of Proserpine. Like others of the ancient oracles, Pythagoras knew by experience that the rumbling echoes of a human voice rolling from a gloomy and sulphurous cavern are irresistibly impressive to an uncritically receptive mind. For descending to such rather shoddy tricks of pedagogy

> Pythagoras has been called a charlatan. He was not. So unquestioning was his belief in his message for his fellow men that he used any and all means within his power to get it accepted. He may even have convinced himself that the voice issuing from the cave was not his own but Apollo's. If so, it would not be the first time or the last that a great teacher elected himself the mouthpiece of divinity. [1, pp. 118–19]

Whatever the truth of Professor Bell's vision, it is the case that the Pythagoreans did make discoveries and that some of them were mathematical. The cult was not one for the masses. It seems to have attracted, as some modern cults have, people with good minds who then used them for the greater glory of Pythagoreanism. The story that the Pythagoreans discovered that harmonious tones are produced when their frequencies have ratios that are small integers is beyond a doubt true, and that the discovery would lead to more investigation of numbers is clear. We do not have to believe Professor Bell's version of the discovery—in fact, we should not—but here it is, for what it is worth.

> Passing a blacksmith's shop one day Pythagoras was arrested by the clang of the hammers swung by four slaves pounding a piece of red-hot iron in succession. All but one of the hammers clanged in harmony. Investigating, he found that the differences in pitch of the four sounds were due to corresponding differences in the weights of the hammers. Without much difficulty he persuaded the blacksmith to lend him the hammers for two hours. In that brief time he was to deflect the course of western civilization toward a new and unimagined goal. With the hammers over his shoulders, he hurried back to Milo's house. There, to the fearful astonishment of the bewildered athlete and his wife, he immediately prepared the first recorded deliberately planned scientific experiment in history.
>
> To each of four strings, all of the same length and of the same thickness, he attached one of the hammers. He next weighed each of the hammers as accurately as he could. How he did this does not matter; he did it. He then hung up the hammers so that the four strings under tension were all of the same length. On plucking the strings, he observed that the sounds emitted corresponded to those made by the hammer striking the anvil. By sticking a small lump of clay on the hammer responsible for the dissonance, he brought the note emitted by its string into harmony with the other three. The four notes, now perfectly harmonized, trembled forth on the air in a melodious chord.
>
> Pythagoras was even more deeply affected than his awestruck audience of two. For in that mysterious chord he recognized the first celestial notes of the elusive music of the spheres which had haunted his dreams since he was a boy. As he knew the weights of the hammers—they should have been perfect globes of pure gold—he quickly inferred the law of musical intervals. To his astonishment he discovered that musical sounds and whole numbers are simply related—how, is immaterial for the moment. It was a great and

> unprecedented discovery, the first hint that the laws of nature may be written in numbers. [1, pp. 101–2]

Some Pythagorean may have made the discovery, but I doubt that it was Pythagoras. Cult leaders and scientists tend to be different types of people. As the classicist Carl Huffman says,

> it is striking how consistently Pythagoras is associated with the manner in which we live our lives rather than with theoretical knowledge.... Plato places Pythagoras's activity in the private sphere and presents him as the model of the sort of figure who was particularly loved as a teacher and whose followers even now are famous for their Pythagorean way of life. Thus, the early evidence consistently portrays Pythagoras as a figure with an extremely loyal following who can tell you how to live your present life in order to best satisfy the gods and to insure the best fate for your soul in the next life. [3]

What I think happened, in those long-ago days when the world of thought was young and ideas were new, was that different Pythagoreans of a mathematical turn of mind found, at one time or another, that strings whose lengths had the ratio of three to two produced harmony when plucked; that

$$1 + 3 + 5 + \cdots + (2n - 1) = n^2$$

(see Figure 1), and that

$$n^2 + \left(\frac{n^2 - 1}{2}\right)^2 = \left(\frac{n^2 - 1}{2} + 1\right)^2$$

(see Figure 2, and then put $2a + 1 = n^2$); that if A, G, and H are the arithmetic,

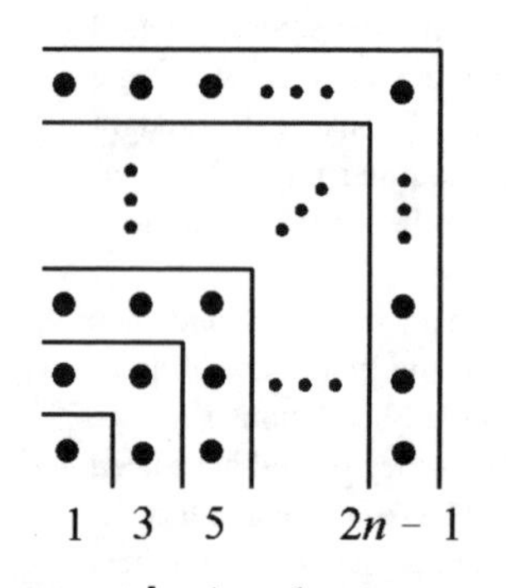

FIGURE 1 An n by n square.

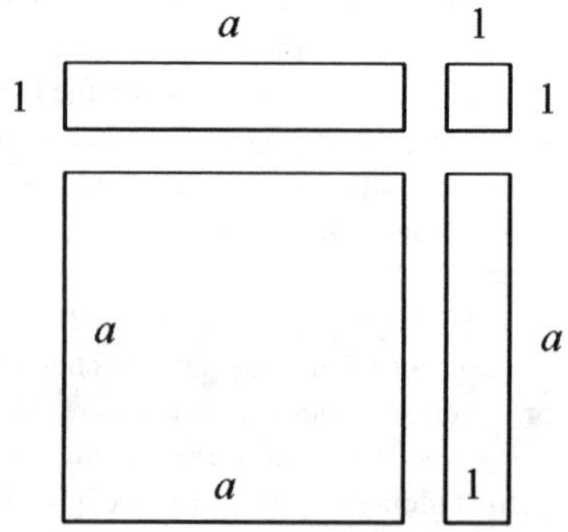

FIGURE 2 An $(a + 1)$ by $(a + 1)$ square.

geometric, and harmonic means of two numbers—

$$A = \frac{a+b}{2}, \quad G = \sqrt{ab}, \quad H = \frac{2ab}{a+b}$$

—then $G = \sqrt{AH}$ is the geometric mean of A and H; that numbers can be classified as deficient, abundant, or perfect by looking at the sums of their divisors,

3	deficient,	$1 < 3$
4	deficient,	$1 + 2 < 4$
5	deficient,	$1 < 5$
6	perfect,	$1 + 2 + 3 = 6$
7	deficient,	$1 < 7$
8	deficient,	$1 + 2 + 4 < 8$
9	deficient,	$1 + 3 < 9$
10	deficient,	$1 + 2 + 5 < 10$
11	deficient,	$1 < 11$
12	abundant,	$1 + 2 + 3 + 4 + 6 > 12$

and so on, through all of the other discoveries attributed to Pythagoras.

Classifying numbers as deficient or abundant is mathematical, but when the Pythagoreans classified them as male or female—odd numbers are male and even numbers female—they were being number-mystical. There were mystical as well as mathematical minds at work among the Pythagoreans. Since ten was for them a holy number ($10 = 1 + 2 + 3 + 4$), it follows (mystically) that there are exactly ten fundamental opposites:

Limited	Unlimited
Odd	Even
One	Many
Right	Left
Masculine	Feminine
Rest	Motion
Straight	Crooked
Light	Darkness
Good	Evil
Square	Oblong

I have no doubt that the Pythagoreans put in as much work on mysticism like this as they did on their mathematics. (It is interesting to see what the Pythagoreans held in low esteem—the items in the second column. Nowadays, we think more highly of motion, multiplicity, and the infinite. Oblongs have also gained in status.)

The two streams of Pythagorean discovery flowed on after the Pythagorean societies themselves died out. Mathematics was taken up and advanced by non-Pythagoreans, and number mysticism also advanced, though we have no classic of number mysticism to compare with Euclid's *Elements*. In fact, number mysticism disappeared from sight, not to resurface for centuries. But it started with the Pythagoreans, and the Pythagoreans started with Pythagoras, so Pythagoras is to blame for number mysticism.

It is easy to see why number mysticism arose. The Pythagoreans discovered amazing things about numbers—about numbers as numbers, about numbers and music, about numbers and geometry, and about numbers and astronomy. As a short biography of Pythagoras says,

> the obscurity concerning Pythagoras's intent has led historians of science into differences of opinion as to whether Pythagoras could really be considered a scientist or even an initiator of scientific ideas. It is further debatable whether those ancient authors who made real contributions to mathematics, astronomy, and the theory of music can be considered to have been true Pythagoreans, or even to have been influenced by authentically Pythagorean ideas. Nevertheless, apart from the theory of metempsychosis (which is mentioned by contemporaries), ancient tradition assigns one doctrine to Pythagoras and the early Pythagoreans that can hardly have failed to influence the development of mathematics. That is the broad generalization, based on rather restricted observation (a procedure common in early Greek science), that all things are numbers. [5, p. 2075]

So, whether or not the discoveries were made by Pythagoreans, the idea that all was number was in their heads. That is a mystical idea, and it is no wonder it had mystical results:

> This speculation about numbers as essences was extended in several directions: as late as the end of the fifth century B.C., philosophers and mathematicians were still seeking the number of justice, or marriage, or even of a specific man or horse. (Attempts were made to discover the number of, for example, a horse, by determining the number of small stones necessary to produce something like the outline of it.) [5, p. 2076]

The fruits of Pythagorean number mysticism are with us today, as will be seen abundantly in later chapters.

I am sure that number mysticism would have appeared even if Pythagoras had never lived. Numbers have power, and they will exert it. As evidence of the power of numbers, notice that there is no such thing as alphabet mysticism even though alphabets go back to before the time of Pythagoras. We have no alphabetologists investigating the mystical properties of "B" or "S," nor do we see those who claim that the end of the world is imminent confirming their ideas alphabetically. The letters of the alphabet are symbols very like the digits of our number system, standing for sound instead of quantity, but they do not inspire mystical insights. Numbers do.

By the way, there is no reason to suppose that Pythagoras, or even the early Pythagoreans, ever proved the Pythagorean theorem. As Huffman says,

> it is clear from the history of Greek mathematics that it is unlikely that Pythagoras could have proved the theorem. A strict proof would require some sort of structure of theorems and definitions like Euclid's *Elements* (I. 47 makes use directly or indirectly of a large number of the preceding 46 theorems) and the first hint of such a structure that we have is connected to Hippocrates of Chios in the later fifth century. We might then suppose that he discovered the theorem in that he recognized its general truth by study of diagrams. However, it is known that the "Pythagorean theorem" had been part of Babylonian arithmetical technique for centuries, although the Babylonians had given no general proof of it. [3]

I should say that my views of Pythagoras, though I think that they are not only correct but obviously so, are not universally agreed with. Leonid Zhmud argues that Pythagoras was in fact a great mathematician and that he made those discoveries that were attributed to him [6]. I am not at all convinced, though Dr. Zhmud does a service by trying to demolish the oft-repeated legend that Pythagoras traveled to Egypt, the Middle East, and even as far as India, absorbing the wisdom of the East, mathematical and otherwise:

> The hypothesis of the Eastern roots of Pythagoras's mathematics is based on the legend of his travels in the East, which has not been confirmed by reliable sources. In addition, the Greek deductive mathematics developed in a plane completely different from that of the calculating mathematics of the Egyptians and Babylonians. No trace of Eastern influence can be found in the works of the Greek mathematicians who had actually visited Egypt, such as Thales, Democritus, and Eudoxus. Even after the conquests of Alexander the Great, when Greeks found themselves living in close contact with these people, they showed no marked tendency to adopt Eastern mathematical methods. [6, p. 250]

One should not be too hard on Pythagoras and his followers. They did make mathematical discoveries, and mathematics is important. It was especially important for the ancient Greeks because it showed them that reason was a way to get at truth. Without that idea there never would have been any philosophy, science, or Western civilization which, despite all their faults, would have been unfortunate.

Dr. Zhmud points this out:

> The first examples of deductive reasoning that we note are the fragments of a philosophical poem by Parmenides and a composition of his disciple Zeno. Parmenides presents his basic tenet: "being is, not-being is not" from which he logically deduces the fundamental signs of being (immutability, unity, unchangeability, etc.) and refutes alternative variants.... Parmenides was perhaps the first *philosopher* to advance his ideas on the fulcrum of logical argumentation, but this is not to say that he invented the method. It is possible to show that he borrowed the deductive method from mathematics and, therefore, it was already in use in the sixth century B.C.
>
> No one will adopt a method if its first application did not give appreciable results in the field in which it arose. It is clear that deductive argumentation in Eleatic philosophy, and in philosophy as such, certainly does not carry the logical conviction and irrefutability of mathematical proof. Neither Parmenides nor Zeno was able to prove anything; they only tried to do so.... When we compare the very minor success of the deductive method in philosophy with what it gave to mathematics, the question of "whom was it borrowed from" seems rhetorical. [6, pp. 251, 253]

So, if Pythagoras was necessary for the progress of mathematics, then he is responsible, indirectly, for a great deal. However, it is possible that mathematics would have developed much as it did even without the Pythagorean impetus. I think that it would have, since it flourished mightily after the Pythagorean influence disappeared. If that is the case, then all that Pythagoras gave the world was number mysticism. Even though it might have surfaced eventually, shame on him.

References

1. Bell, Eric Temple, *The Magic of Numbers*, McGraw-Hill, New York, 1946, reprinted by Dover, New York, 1991.
2. Henderson, Janice A., Pythagoras of Samos, in *The Software Toolworks Multimedia Encyclopedia*, version 1.5, Grolier Electronic Publishing, 1992.
3. Huffman, Carl, Pythagoras and Pythagoreanism, in *Le savoir grec*, edited by G. E. B. Lloyd and J. Braunschvig, Flammarion, Paris, to appear.
4. Portis, Charles, *Masters of Atlantis*, Knopf, New York, 1985.
5. Von Fritz, Kurt, Pythagoras of Samos, in *Biographical Dictionary of Mathematicians*, Scribner's, New York, 1991.
6. Zhmud, Leonid, Pythagoras as a mathematician, *Historia Mathematica* **16** (1989), 249–268.

CHAPTER 3

Neopythagoreanism

Although with the Pythagoreans mathematics and number mysticism went hand-in-hand, the separation of the two was probably not long in coming. Certainly there is no trace of mysticism in Euclid's *Elements*. There are many traces of it in the works of Plato, but though he was a great fan of mathematics, he knew it from the outside and so there were large gaps in his knowledge. The space thus unoccupied was available to be filled with mysticism. After Plato, number mysticism starts to disappear, or to go underground. Though we have works of mathematicians in the last centuries B.C., we lack corresponding works of number mystics.

There is a gap in the written record of number mysticism of hundreds of years, until Nicomachus's *Arithmetic* at the end of the first century A.D. A natural question is, what was going on in those hundreds of years? Was there an active oral tradition, continually developing, whose written counterpart has been lost? That could have been, but I do not think so. What follows is speculation only, with no evidence behind it other than intuition. In fact, it could be called history mysticism.

I think that Pythagoreanism essentially disappeared after the dissolution of the Pythagorean cult groups. Mathematicians certainly paid no attention to it. Anyone with an inclination toward mathematics and who had the necessary intellectual power would prefer studying Apollonius's magnificent work on the conic sections to contemplating the fog of number mysticism, and though Pythagorean thought was not completely forgotten, I think it was disregarded. Pythagoreanism was a product of the early days, not only the early days of mathematics but the early days of thought. Those of us who come later look back on the early days with admiration for the accomplishments of the pioneers, but also with satisfaction that we have progressed beyond their errors and false starts.

The fifth, fourth, and third centuries B.C. were a time of great advances for Greek culture on all fronts. The Parthenon was built, Aristotle was founding physics and Hippocrates medicine, literature flourished, *progress* was being made on all fronts. With so much else to do, mysticism could wait. There were more important things to be done.

Number mysticism waited until progress slowed, stopped, and started to go into reverse. The Romans came a-conquering. The pace of mathematical discovery slowed, and commentaries on the writings of the masters of the past replaced new discoveries. The golden age was gone, the glory days were past. At such times, people's eyes turn inward. When external reality no longer brings rewards, then we remember that it is internal reality that really matters. Striving after material goods, building bigger and better buildings, extending the enlightenment of civilization to those in darkness, proving a new theorem—why bother with ephemera like those when only spiritual truth is eternal? I think that it was the decline of Greek civilization that caused Pythagoreanism to be rediscovered, centuries after Pythagoras, and given the impetus that it has had to this day.

Whether the current popularity of mysticism, numerical and otherwise, is evidence that our civilization is sliding downhill remains to be seen. At some time in the future, hindsight will clearly show what we should have seen all along but did not.

Whatever the causes of the revival of Pythagoreanism, a revival there was, and it seems likely that it was due to one person. Robin Waterfield, the translator of Iamblichus's *The Theology of Arithmetic*, says that

> more importantly, the later extant texts show a high degree of unanimity of thought and even of language. This suggests that there was some common seminal writer, whose name is unknown to us (it used to be thought to be the polymath Posidonius (c. 135–c. 50 B.C.)), but who can be dated to the second century B.C. [2, p. 27]

Similarly, as we will see in chapter 19, Mrs. L. Dow Balliett, whatever the source of her ideas, singlehandedly put modern numerology on the map. Writers after her also show a high degree of unanimity.

One person can have a large influence:

> This unknown teacher seems to have given a new impetus to Pythagoreanism. The name of Pythagoras was no doubt never far from the lips of all the arithmologists of these centuries; but in the first century B.C., in Rome and Alexandria, thinkers again began to claim direct descent from the master, and we nowadays call them Neopythagoreans.... the early Neopythagoreans also paved the way for Greek arithmology to enter the Jewish tradition via the works of Philo of Alexandria, and the Christian tradition via the works of Clement of Alexandria. [2, p. 28]

This unknown, who was to Pythagoras as Paul was to Jesus (though with more time in between) has a lot—all of modern numerology—to answer for.

Another unanswerable question is, what would have become of Pythagoreanism if the unknown had not done his unknown-to-us work? Would Pythagoreanism have surfaced through some other means?

You could answer, no, it wouldn't have. Without the unknown, Pythagoreanism might have disappeared as have innumerable other cults through the centuries. Something else could have arisen to occupy those minds susceptible to such thoughts as are found in Pythagoreanism, but there is no reason to think that it would have any relation to the Pythagoreans.

Or you could maintain that, yes, it would have. The power of numbers is not to be denied, and it will be exerted through one medium or another. The extent of the power is shown by the early appearance of thought about numbers in the history of human thought. People *will* think about numbers. Just as a river, no matter how dammed, will find its way to the sea, so it may be that the power of numbers, including the power of number mysticism, will be not denied. As I said, the question is unanswerable, but I think that neo-Pythagoreanism was inevitable.

Whatever their cause, we have the works of the Neopythagoreans. I will quote from [2] to show the state of "NP" thought some 1500 years ago. This will also serve as an introduction to number mysticism for those unfamiliar with the field. As you will see, NP thought is still number mysticism and not yet numerology—that is, it is concerned with mystical appreciations of the properties of integers, and the properties have no application to anything other than mysticism. The applications came later. However, modern numerologists ascribe the same properties to integers as Iamblichus did. Either mystical truths are eternal or numerological writers copy from each other.

The Theology of Arithmetic may not actually have been written by Iamblichus, but that really is no matter:

> The treatise is, in fact, a compilation and reads like a student's written-up notes. Whole sections are taken from the *Theology of Arithmetic* of the famous and influential mathematician and philosopher Nicomachus of Gerasa, and from *On the Decad* of Iamblichus' teacher, Anatolius, Bishop of Laodicea. These two sources, which occupy the majority of our treatise, are linked by text whose origin is at best conjectural, but some of which could very well be lecture notes—perhaps even from lectures delivered by Iamblichus. At any rate, the treatise may tentatively be dated to the middle of the fourth century A.D. [2, p. 23]

Just as most modern books on numerology, the *Theology* gives properties of the digits, one by one. Unlike modern books, it supposes some knowledge of Pythagorean mathematics. This is from Iamblichus, or his notetaker:

> The monad is the non-spatial source of number. It is called "monad" because of its stability, since it preserves the specific identity of any number with which it is conjoined. For instance, $3 \times 1 = 3$, $4 \times 1 = 4$: see how the approach of the monad to these numbers preserved the same identity and did not produce a different number.
>
> Everything has been organized by the monad, because it contains everything potentially: for even if they are not yet actual, nevertheless the monad holds seminally the principles which are within all numbers, including those which are within the dyad. For the monad is even and odd and even-odd; linear and plane and solid (cubical and spherical and in the form of pyramids from those with four angles to those with an indefinite number of angles); perfect and over-perfect and defective; proportionate and harmonic; prime and incomposite, and secondary; diagonal and side; and it is the source of every relation. [2, p. 35]

Readers were assumed to be familiar with the Pythagorean classifications of number—even, odd, perfect, prime, and so on. It may seem strange that the monad should be both even and odd, but that is because some properties hold vacuously. The monad is even because it is not odd, since what makes an integer odd is its divisibility into two equal parts with one left over. Since the monad cannot be divided, it cannot be odd. The monad is odd because it is not even, since even numbers are those that can be divided into two equal parts. For the monad there is no question of this since it is indivisible. If you object that every integer must be even or odd, you have not yet progressed far into number mysticism. The monad does not have to be even or odd because the monad is not a number. It is the monad.

The monad is linear, plane, and solid because it is a first power, a square, and a cube. It is prime because it is not composite, once again because of indivisibility, and it is secondary because it is not prime since the primes are a subset of $\{2, 3, 5, 7, \ldots\}$ and 1 is not included. I am not sure how it can be simultaneously perfect, abundant ("over-perfect"), and deficient ("defective") all at once; it could be because those adjectives are descriptive of number and the monad, though one, is not one. (Perhaps that pun should be disregarded. Number mysticism is *serious*.) In any event, the NPs knew a good deal more about mathematics than do their modern successors, the numerologists. Modern numerologists would not know an abundant number if it hit them.

Iamblichus then moves away from mathematics into the mystical associations of the monad:

> Just as without the monad there is in general no composition of anything, so also without it there is no knowledge of anything whatsoever, since it is a pure light, most authoritative over everything in general, and it is sun-like and ruling, so that in each of these respects it resembles God, and especially because it has the power of making things cohere and combine, even when

> they are composed of many ingredients and are very different from one another, just as he made this universe harmonious and unified out of things which are likewise opposed.
>
> Furthermore, the monad produces itself and is produced from itself, since it is self-sufficient and has no power set over it and is everlasting; and it is evidently the cause of permanence, just as God is thought to be the cause of actual physical things, and to be the preserver and maintainer of natures. [2, p. 37]

Notice that the monad is the cause of permanence, on a par with God, the cause of things. Oh, the power of numbers!

Gematria existed:

> The mark which signifies the monad is a symbol of the source of all things. And it reveals its kinship with the sun in the summation of its name: for the word "monad" when added up yields 361, which are the degrees of the zodiacal circle. [2, p. 39]

Gematria is the art and science of assigning numbers to letters and thus to words, the better to understand their mystical significance. It arose because the ancient Greeks used the letters of their alphabet to represent integers:

α	β	γ	δ	ε	ϛ	ζ	η	θ
1	2	3	4	5	6	7	8	9
ι	κ	λ	μ	ν	ξ	ο	π	ϙ
10	20	30	40	50	60	70	80	90
ρ	σ	τ	υ	φ	χ	ψ	ω	ϡ
100	200	300	400	500	600	700	800	900

6, 90 and 900 were represented as numbers with obsolete letters that did not appear in the language and so could not enter into gematria: ϛ (digamma, or vau) for 6, ϙ (koppa) for 90, and ϡ (sampi) for 900.

Sure enough,

$$\mu + o + \nu + \alpha + \varsigma = 40 + 70 + 50 + 1 + 200 = 361.$$

There are only 360 degrees in the circle, but

> Presumably 361, rather than 360, is given as the number of degrees because the first one is counted twice, to indicate a complete circle. [2, p. 31n]

The essence of number mysticism is giving nonnumerical associations to numbers:

> The Pythagoreans called the monad "intellect" because they thought that intellect was akin to the One; for among the virtues, they likened the monad to moral wisdom; for what is correct is one. And they called it "being," "cause

> of truth," "simple," "paradigm," "order," "concord," "what is equal among greater and lesser," "the mean between intensity and slackness," "moderation in plurality," "the instant now in time," and moreover they called it "ship," "chariot," "friend," "life," "happiness." [2, p. 39–40]

In any event, it is One, it is Origin and it is anything that mystically follows from those.

Two, the dyad, like the monad, is not a number:

> The dyad is not number, nor even, because it is not actual; at any rate, every even number is divisible into both even and unequal parts, but the dyad alone cannot be divided into unequal parts; and also, when it is divided into equal parts, it is completely unclear to which class its parts belong, as it is like a source. [2, p. 45]

Every even number, except 2, can be divided in more than one way—for example, $4 = 2 + 2 = 3 + 1$—2 is special. When other integers are divided into parts, as $7 = 4 + 3$, it is clear to what classes its parts belong, but when the dyad is divided into two parts, two monads result. But, as we have seen, the monad belongs in *all* classses: even, odd, prime, composite, and so on. Thus the dyad cannot be a number.

Further evidence that the dyad is not a number is that

> the dyad would be the midpoint between plurality, which is regarded as falling under the triad, and that which is opposed to plurality, which falls under the monad. Hence it simultaneously has the properties of both. It is the property of 1, as source, to make something more by addition than by the blending power of multiplication (and that is why $1 + 1$ is more than 1×1), and it is the property of plurality, on the other hand, as product, to do the opposite: for it makes something more by multiplication than by addition. For plurality is no longer like a source, but each number is generated out of another and by blending (and that is why 3×3 is more than $3 + 3$). And while the monad and the triad have opposite properties, the dyad is, as it were, the mean, and will admit the properties of both at once, as it occupies the mid-point between each. [2, pp. 43–44]

That is,

$$1 \times 1 < 1 + 1$$

$$2 \times 2 = 2 + 2$$

$$n \times n > n + n \quad \text{for } n \geq 3$$

showing the special place of the monad and the dyad apart from the numbers.

Similarly, polygons all have at least three sides, showing that numbers begin at three, and also allowing for a mystical interpretation:

> The dyad is clearly formless, because the infinite sequence of polygons arise in actuality from triangularity and the triad, while as a result of the monad everything is together in potential, and no rectilinear figure consists of two straight lines or two angles. So what is indefinite and formless falls under the dyad alone. [2, p. 45]

The dyad is the moon:

> And they say that the name "dyad" is suited to the moon, both because it admits of more settings than any of the other planets, and because the moon is halved or divided into two. [2, p. 47]

True enough, the moon periodically displays half of itself, but it is hardly clear what its settings have to do with its two-ness. This is because the passage has been translated from Greek into English and the pun—*duas* for dyad and *duseis* for settings—is lost. Language as well as number can generate mystical insights.

Naturally, the dyad is sex:

> The dyad, they say, is also called "Erato"; for having attracted through love the advance of the monad as form, it generates the rest of the results, starting with the triad and tetrad. [2, p. 46]

It is, in a sense, evil:

> So the dyad alone remains without form and without the limitation of being contained by three terms and proportionality, and is opposed and contrary to the monad beyond all other numerical terms (as matter is contrary to God, or body to incorporeality), and as it were the source and foundation of the diversity of numbers, and hence resembles matter; and the dyad is all but contrasted to the nature of God in the sense that it is considered to be the cause of things changing and altering, while God is the cause of sameness and unchanging stability. [2, p. 41]

With three we at last break into the realm of true number.

> The monad is like a seed in containing in itself the unformed and also unarticulated principle of every number; the dyad is a small advance towards number, but is not number outright because it is like a source; but the triad causes the potential of the monad to advance into actuality and extension. [2, p. 50]

We also have the first glimmerings of the very important Law of Small Numbers (which states that there are not enough small integers to carry out the many tasks assigned to them—see chapter 9):

> "This" belongs to the monad, "either" to the dyad, and "each" and "every" to the triad. Hence we use the triad also for the manifestation of plurality, and

> say "thrice ten thousand" when we mean "many times many," and "thrice blessed." Hence too we traditionally invoke the dead three times. Moreover, anything in Nature which has process has three boundaries (beginning, peak and end—that is, its limits and its middle), and two intervals (that is, increase and decrease), with the consequence that the nature of the dyad and "either" manifests in the triad by means of its limits. [2, pp. 50–51]

The idea of beginning, middle, and end has mystical extensions:

> The triad is called "prudence" and "wisdom"—that is, when people act correctly as regards the present, look ahead to the future, and gain experience from what has already happened in the past: so wisdom surveys the three parts of time, and consequently knowledge falls under the triad....
>
> They call it "friendship" and "peace," and further "harmony" and "unanimity": for these are all cohesive and unifactory of opposites and dissimilars. Hence they also call it "marriage." And there are three ages in life. [2, pp. 51, 53]

Another Greek source claims that there are four ages in life: childhood, youth, maturity, and old age, all equally long. Shakespeare said that there are seven. If you agree with those who say that the atoms of the body completely change every seven years, then there are ten. Thus does numerology give flexibility: you can choose whichever of 3, 4, 7, and 10 is most convenient.

Four is the number of knowledge:

> ... the monad of arithmetic, the dyad of music, the triad of geometry, and the tetrad of astronomy, just as in the text entitled *On the Gods* Pythagoras distinguishes them as follows: "Four are the foundations of wisdom—arithmetic, music, geometry, astronomy—ordered 1, 2, 3, 4." [2, p. 56]

Arithmetic comes first because it is the monad, from which all else proceeds:

> ... for when arithmetic is abolished, so are the other sciences, and they are generated when it is generated, but not vice versa, with the result that it is more primal than them and is their mother, just as the monad evidently is as regards the numbers which follow it. [2, p. 56]

Music is a dyad because it

> obviously pertains to difference in some way, since it is a relation and a harmonious fitting together of things which are altogether dissimilar and involved in difference. [2, p. 56]

Geometry is three because there are three dimensions;

> and astronomy—the science of the heavenly spheres—falls under the tetrad, because of all solids the one most perfect and the one which particularly embraces the rest by nature, and is outstanding in thousands of other aspects, is the sphere, which is a body consisting of four things—center, diameter, circumference and area. [2, p. 57]

If a number mystic wished to associate spheres with five instead of four, he could just as well say that spheres consist of five things—center, diameter, circumference, area, and volume. Mysticism is different from deduction. In deduction, the reasons determine the conclusion, but in mysticism the conclusion can determine the reasons, or at least influence them strongly.

Four is the number of the world, since there are four elements—fire, air, water, and earth—four seasons, four senses ("touch is a common background to the other four, which is why it alone does not have a location or a regular organ" [2, p. 59]), four parts of the body—head, trunk, legs, and arms—four winds, and so on through other applications of the Law of Small Numbers.

The Law operates also for five: there are five planets (Mercury, Venus, Mars, Saturn, and Jupiter), five regular solids (tetrahedron, cube, octahedron, dodecahedron, and icosahedron), plants have five parts (root, stem, bark, leaf, and fruit), there are five precipitations (rain, snow, dew, hail, and frost), five circles on the earth (the equator, the tropics of Cancer and Capricorn, and the Arctic and Antarctic circles), five genera of creatures (those that live in fire, in air, on earth, in water, and the amphibians), and so on.

To test the universality of number mysticism you could take a survey and ask people how many parts plants had, or how many kinds of precipitations there are, and see if everyone, or even a majority, said that there were five. There would be disagreements. Mysticism is personal and not duplicatible. It is not scientific. But then if it were scientific, it would be a science.

Five is preeminently the number of justice, because it is in the center of the digits

$$1 \quad 2 \quad 3 \quad 4 \quad 5 \quad 6 \quad 7 \quad 8 \quad 9$$

> So, you see, the pentad is another thing which has neither excess nor defectiveness in it, and will turn out to provide this property for the rest of the numbers, so that it is a kind of justice, on the analogy of a weighing instrument. For if we suppose that the row of numbers is some such weighing instrument, and the mean number 5 is the hole of the balance, then all the parts towards the ennead, starting with the hexad, will sink down because of their quantity, and those toward the monad, starting with the tetrad, will rise up because of their fewness, and the ones which have the advantage will altogether be triple the total of the ones over which they have the advantage, but 5 itself, as the hole in the beam, partakes of neither, but it alone has equality and sameness. [2, p. 70]

It is too bad that we do not have six fingers on each hand. Then we would count by dozens and there would be eleven digits instead of nine,

$$1 \quad 2 \quad 3 \quad 4 \quad 5 \quad 6 \quad 7 \quad 8 \quad 9 \quad \chi \quad \varepsilon$$

the last two ("dek" and "el") for our two extra fingers. Then 6, a perfect number and the center of the digits would, fittingly, be the number of justice. On the other hand, numerology books, which go on at great length about the properties of each digit, would all be longer by a factor of $11/9$.

At the end of the section on the pentad, Iamblichus (or the student taking the lecture notes) could not resist including Nicomachus's solution of the problem of the existence of evil. It has nothing to do with the pentad, except for the connection between evil and justice, and is probably there just because he was so impressed by it. Here it is:

> When men are wronged, they want the gods to exist, but when they commit wrong, they do not want the gods to exist; hence they are wronged so that they may want the gods to exist. For if they do not want the gods to exist, they do not persevere. Therefore, if the cause of men's perseverance is their wanting the gods to exist, and if they want gods to exist when they are wronged, and if wrong, though an evil, still looks to Nature's advantage, and whatever looks to Nature's advantage is good, and Nature is good, and Providence is the same, then evil happens to men by Providence. [2, p. 74]

Is that not good? The ancient Greeks, even the number mystics among them, had good heads on their shoulders.

Six is the first perfect number, though the first Pythagoreans did not consider such things. They did, however, think highly of six, as the following shows. It also shows some of the Pythagorean attitude towards number: matter *wants* to be number.

> After the pentad, they used naturally to praise the number 6 in very vivid eulogies, concluding from unequivocal evidence that the universe is ensouled and harmonized by it and, thanks to it, comes by both wholeness and permanence, and perfect health, as regards both living creatures and plants in their intercourse and increase, and beauty and excellence, and so on and so forth.

(That last phrase could well have been written by a notetaker, impatient with the lecturer's prolixity.)

> They undertook to prove this by adducing the following as evidence: the disorder and formlessness of the eternal prime matter, and lack of absolutely everything which makes for distinctness, was separated out and made orderly by number, since number is the most authoritative and creative kind of thing, and matter in fact partakes of distinctness and regulated alteration and pure coherence thanks to its desire for and imitation of the properties of number.
> [2, p. 76]

The reason for the specialness of 6 is that primary perfection consists of having a beginning, middle, and end, and if the integers are divided into triads, the triads are "given their identity" by the hexad, since

$$\begin{array}{rl} 1+2+3=6 & \\ 4+5+6=15 & 1+5=6 \\ 7+8+9=24 & 2+4=6 \\ 10+11+12=33 & 3+3=6 \end{array}$$

and so on.

When we come to seven, we have a mystery. The first sentence in its section is

> Seven is not born of any mother and is a virgin. [2, p. 87]

It is natural to expect a little explanation, but there is none. The next sentence is

> The sequence from the monad to it added together totals 28; the 28 days of the moon are fulfilled hebdomad by hebdomad.

After that Iamblichus, or his notetaking student, swings into usual things like

> We are seven things—body, distance, shape, size, color, movement and rest. There are seven movements—up, down, forward, backward, right, left and circular. Plato composed the soul out of seven numbers. Everything is fond of sevens. There are 7 ages, as Hippocrates says...

and so on.

There is a passage later that touches on the isolation of seven, but without shedding very much light:

> Moreover, the hebdomad seems to be an acropolis, as it were, and a "strong fortification" within the decad, just like an indivisible monad. For it alone admits no breadth, since it is a rectilinear number and admits only a fractional part with the same denominator as itself, and, by mingling with any of the numbers within the decad, it does not produce any of the numbers within the decad, nor is it produced by the intercourse of any of the numbers within the decad, but, with a principle which is all its own and is not shared, it has been assigned the most critical place. [2, pp. 89-90]

What in the world could that mean?

> They called the heptad "Athena" and "critical time" and "Chance"—"Athena" because it is a virgin and unwed, just like Athena in myth, and is born neither of mother (i.e. of even number) nor of father (i.e. odd number), but from the head of the father of all (i.e. from the monad, the head of number); and like Athena it is not womanish, but divisible number is female. [2, p. 99]

What could *that* mean?

The problem is too much for me, but a plausible answer was given in 1936 by Grace Murray Hopper [1]. (Before she was an admiral and before she

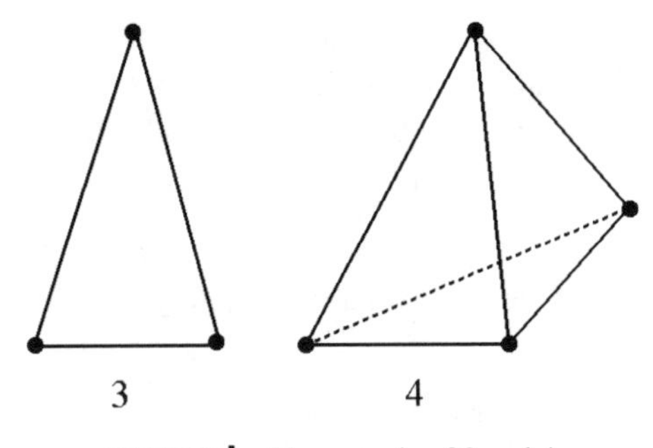

FIGURE 1 The genesis of 3 and 4.

was a computer expert, she was a classical scholar.) Iamblichus says that seven was not produced by any of the numbers within the decad. Admiral Hopper explained that geometrically. She said that the other numbers arise naturally, and seven is left out. So, as in Figure 1, three is the first plane number and four is the first solid number. Then three and four generate five and six in the 3-4-5 right triangle, with hypotenuse 5 and area 6 (Figure 2). Eight is the second solid number and nine (Figure 3) is the third square number. What the ancient Greek number mystics had in mind when they spoke of the spontaneous generation and virginity of seven we will never be sure, but Admiral Hopper's hypothesis is sufficiently satisfying for me.

By the way, many writers on numerology assert that five is the number of marriage because it is the sum of two and three, the first female and male numbers. This is an error, for two reasons: first, two is the dyad and is not a number and second, the male number must come first. The 3-4-5 triangle provides a better explanation: 3, the first male number and 4, the first female number, combine to produce 5.

Many of the properties of seven had to do with bodies, and life.

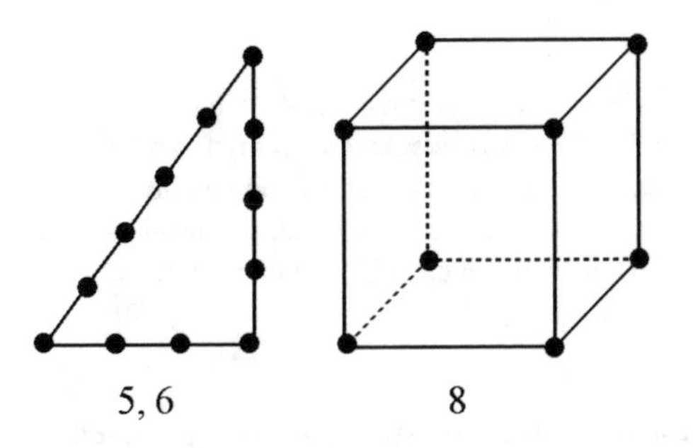

FIGURE 2 The birth of 5, 6, and 8.

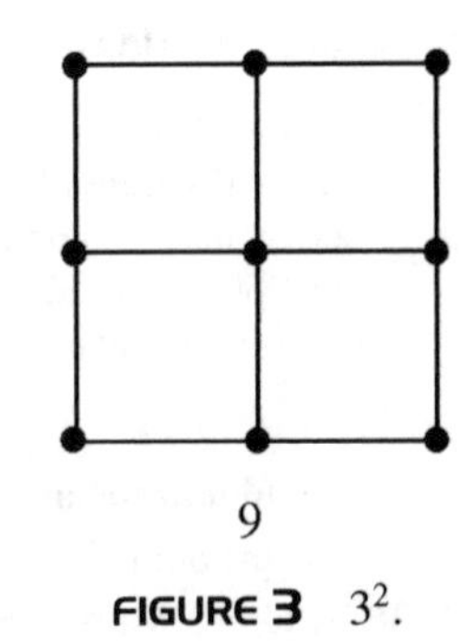

FIGURE 3 3^2.

> All seeds appear above ground, during growth, in the course of the seventh day or thereabouts, and the majority of them are seven-stemmed for the most part. Just as foetuses were sown and ordered in the womb by the hebdomad, so after birth in seven hours they reach the crisis of whether or not they will live. For all those which are born complete and not dead come out of the womb breathing, but as regards the acceptance of the air which is being breathed and by which soul in general acquires tension, they are confirmed at the critical seventh hour one way or the other—either towards life or towards death. [2, p. 94]

Now we are up to eight—the ogdoad—and either Iamblichus or his notetaker is running out of steam. Where 4 had nine pages of text devoted to it and 5 had ten, 8 rates only four and 9 has only three. 8 has a few arithmetical properties, such as being $5+3$, more elementary than the next cube, $27 = 7+9+11$, or the one after that, $64 = 13 + 15 + 17 + 19$; a few physical properties, such as the eight spheres of the heavenly bodies; and a few mystical properties:

> Philolaus says that after mathematical magnitude has become three-dimensional thanks to the tetrad, there is the quality and "color" of visible nature in the pentad, and emolument in the hexad, and intelligence and health and what he calls "light" in the hebdomad, and then next, with the ogdoad, things come by love and friendship and wisdom and creative thought. [2, p. 103]

And so we end with nine:

> At any rate, because it does not allow the harmony of number to be dissipated beyond itself, but brings numbers together and makes them play in concert, it is called "concord" and "limitation," and also "sun," in the sense that it gathers things together.
>
> It is called "lack of strife" because of the correspondence and interchange of numbers from it to the monad, as was discussed in the diagram about justice.
>
> It was called "assimilation," perhaps because it is the first odd square (for odd numbers are called "assimilative" in general because of assimilation; and moreover, squares are assimilative, oblongs are dissimilar), and perhaps also because it is particularly assimilated to its square root. [2, p. 106]

The time has come to let the notetaker alone, to prepare for the final examination.

Neopythagoreanism differs from Pythagoreanism in that numbers are both less and more. Less, since number is no longer *all* (assuming that the Pythagoreans in fact believed that), but more because numbers have, in their assimilative fashion, taken on properties that the Pythagoreans did not give them.

After the Greek Neopythagoreans, number mysticism seems to have once again subsided. The practical Romans did not spend any time on it. It survived into the middle ages in part because Boethius summarized the *Arithmetic* of

Nicomachus which thus became part of the curriculum of higher education for hundreds of years. We will later see what form numerology and number mysticism took during the middle ages, and how, one hundred years ago, Mrs. L. Dow Balliett founded modern numerology.

References

1. Hopper, Grace Murray, The ungenerated seven as an index to Pythagorean number theory, *American Mathematical Monthly* **43** (1936) #7, 409-413.
2. Iamblichus, *The Theology of Arithmetic*, translated by Robin Waterfield, Phanes Press, Grand Rapids, Michigan, 1988.

CHAPTER 4
The Pythagoreans Abroad

An idea like Pythagoreanism is hard to kill. It had a resurgence about fifty years ago in the American Institute of Man, whose fantastic theories show, once again, the power that numbers have, in some minds, over reality.

AIM was incorporated in the state of Illinois as a not-for-profit scientific research organization in 1944. Its founder and president, Alexander Ebin, said that

> the Institute was founded after 18 years of preliminary research disclosed that a large part of the history of Man has not been accurately reported....
>
> The Institute is based explicitly upon the tradition of Pythagoras and his school. Its corporate seal is the identification sign used by the "Mathematicians" of Croton: the pentagon-star inscribed in the circle.
>
> There are two classes of members, corresponding to the two membership groups (Hearers and Mathematicians) of the first Pythagorean society. *Ordinary members* pay annual membership dues and receive the Institute's publications. *Creative members* have no financial obligation and contribute creatively to the Institute's work....
>
> *Ordinary Members:* Any interested person may become an ordinary member of the American Institute of Man simply by expressing a desire for membership and by paying five dollars with application for one year's membership dues. One-half of this amount will be used to support the Institute's scientific work, and one-half will be used to inform and entertain the member.
>
> *Creative Members:* Creative membership is by invitation only, and is subject to Pythagorean rules. Each creative member agrees to serve a five-year probationary period before his membership becomes final, and to waive personal and public credit for his creative contributions to the work of the Institute during his lifetime. (His creative contributions will be publicly acknowledged at his death.) Any creative member who claims public credit for his creative contributions is automatically expelled from membership and the Institute will not, during his lifetime or after his death, acknowledge his work. [1, p. i]

Thought had clearly gone into the organization of AIM. The publication listed a President, Vice-President and Secretary, but gave no membership figures. I would not be surprised to find that its membership never consisted of many more than its founder and a few of his friends. All I know about the founder is that he wrote in 1966 that he had been a professional editor for more than forty years, for part of that time with the *New York Times*. His prose style is that of a practiced writer. In 1966 he was continuing to send his startling Pythagorean discoveries to the Rockefeller Foundation and the Carnegie Institution, probably to no effect.

His ideas about Pythagoras were not startling, though not all are supported by historical evidence:

> Pure mathematics was invented by definition, from pre-existing mathematical ideas, by Pythagoras of Samos. In his time, the 6th century B.C., there did not exist in the world a universal religion. There were various national religions, expressed in national languages and symbols, which included elements of science. For example, the national religion of Egypt included Egyptian science as part of its mystery. To learn Egyptian science it was first necessary to learn the Egyptian language and then to gain permission, a matter of some difficulty, to study the Egyptian mysteries. Pythagoras was able to secure this permission because of the friendship between his patron, Polycrates of Samos, and King Amasis of Egypt. It is reported that Pythagoras also studied the Orphic mysteries with Pherecydes of Syros and the mysteries of Chaldea. After completing his studies of various national religions Pythagoras returned to Samos with the idea of creating the world's first universal religion, i.e., a religion, embracing all knowledge, which would meet the need of all races and nations of mankind: a veritable religion of Man. His first problem, made vividly clear to him by his studies, was to create a universal language through which men of all nations could understand each other and reach enduring agreements about questions of common knowledge. He had observed in his travels that the elements for such a language existed, ready to hand, in the common language of the market-place. Men who could not speak each other's languages or share each other's customs, could nevertheless trade equitably with one another by counting, weighing, and measuring. He proposed to take this commercial language, purify it of its commercial associations, and create a universal language to serve as a vehicle for a universal religion. Thus there came into being the new language of "pure mathematics" which, after nearly 25 centuries, has proved its worth as a universal language. [1, pp. 2–3]

One of the difficulties with this is that Thales, born 70 or so years before Pythagoras, is given credit for proving the first theorem—that is, for making the momentous discovery that mathematical truths could be deduced rather than observed—and this could be taken as marking the start of pure mathematics. Another is the tales about travel and study. Similar stories are told about

Thales, with the same lack of basis. The final difficulty is asserting knowledge of Pythagoras's motives. However, AIM's interpretation of history does not depart very much, so far, from what everyone thinks.

Soon, though, it takes off. The general view is that the Pythagorean cult, subject to occasional persecution, disappeared in the fourth century B.C. and that Pythagoreanism was quiescent until its Neopythagorean revival in the second century A.D. AIM had additional information.

> As a result of reconstructing the ideas and history of the Pythagoreans, we have identified and partly deciphered a large amount of heretofore unsuspected documentation for the early history of pure and applied mathematics....
>
> It is not correct to say that we have "discovered" new documentation for the history of mathematics because our materials are "well-known" to different groups of non-mathematical specialists: to classical scholars, archeologists, anthropologists, etc.... Our position is that by learning to read the special idiom of Pythagorean mathematics we have been able to recognize in the materials of history, and to read in part, a large number of Pythagorean records which were left in public archives between the first and seventh centuries A.D.... This action opens up a controversial situation with which we have been contending privately for five years. As the Pythagorean records are identified and deciphered they will have a revolutionary effect upon fields of knowledge that are now remote from mathematics. In particular, anthropologists and archeologists will have to revise radically some of their cherished notions about the history of civilization and the development of specific cultures.... Changes of this magnitude will not be accomplished without resistance from specialists in the subjects affected. The resistance may take the form of denying the validity of mathematical proofs in the historical field, even when the historical materials are mathematical and refer to the history of mathematics. [1, pp. 1–2]

We will soon see examples of AIM's "proofs" and they will explain the resistance to—"dismissal" or "ignoring of" would be more accurate—their conclusions. AIM claims that previously unknown historical events must have occurred because of some properties of numbers. This is numerology: when it comes to a choice between historical evidence and what some numbers seem to say, the numbers win.

According to AIM, the Pythagorean society did not disappear, it moved west to Massalia (Marseilles).

> About 330 B.C. the Pythagoreans of Massalia initiated a large-scale program of geographical exploration and missionary activity. Pytheas of Massalia developed the first true method for measuring geographical locations, equivalent to the measurement of latitude, and organized an expedition to explore and measure the coastline of western Europe and the islands of the North Atlantic Ocean.

He got as far as Iceland.

> From this it follows that Pytheas discovered North America at Iceland more than 18 centuries before Columbus discovered North America at Watling Island in the Bahamas in 1492. It is relevant to our announcement to note that North America was first discovered, for science, by a Pythagorean expedition because the largest number of Pythagorean records, including hundreds of dated mathematical inscriptions carved in stone, were left on public record in ancient America. Some of these records commemorate Pytheas of Massalia, by name and place, as the discoverer both of Iceland and of America.

The druids were Pythagoreans:

> During the course of his expedition, Pytheas spent some time in Britain where he founded a Celtic-Pythagorean school, known by the Greco-Celtic name *Druid*, which flourished and maintained cultural relations with Massalia until the Roman invasion and conquest of Britain in A.D. 43–61. The history and accomplishments of this school are beautifully commemorated by stone monuments which still stand, unrecognized, in Britain. [1, p. 4]

Other Pythagoreans went east instead of west, where they invented the telescope and calculus:

> In parallel with the voyage of Pytheas, between 331 and 323 B.C., Pythagorean ideas entered Persia and India with Alexander of Macedon who, as part of his program for hellenizing Asia, founded many Greek cities and colonized them with educated Greeks. In India, Pythagoreanism merged with Buddhism to create a new Greco-Buddhist religion with a distinctive art (Gandhara) and theology (Mahayana) and a Pythagorean interest in mathematics. This Greco-Indian school developed certain ideas which are basic to modern mathematics, notably the concept of zero and arithmetical algebra.
> [1, pp. 4-5]

It is true that zero and a rudimentary algebra developed in India, but no one else thinks that happened until at least 300 A.D.

> By the early part of the first century A.D. an advanced optical theory had been developed and efficient telescopes invented. Telescopic study of the planet Venus, which exhibits phases like the Moon, led to a formulation of a heliocentric theory of the solar system. To deal with the problem of planetary motion an efficient method was devised for computing rates of change. Following the line of progress initiated by Hipparchus, who founded the science of measuring longitude, an efficient lunar theory was developed. The foregoing advances culminated, about A.D. 30–40, in the first efficient science of navigation. Partly to test this theory an expedition was organized to circumnavigate the earth by sailing eastward from the mouth of the Indus River. This expedition left the island of Hainan in A.D. 42–43 and reached South America, via the islands of the South Pacific Ocean, about A.D. 43. Thus, South America was discovered independently of North America, by

> an expedition from India, about 370 years after the voyage of Pytheas and more than 1400 years before the voyage of Columbus.
>
> This Indian expedition explored the Pacific and Atlantic coastlines of South, Middle, and North America and, to verify Pytheas' account of his discovery of "Thule" under the Arctic Circle, sailed eastward from North America to Greenland and thence to Iceland. The expedition reached Britain from Iceland at some undetermined time after the Roman invasion of A.D. 43. British Pythagoreans joined the expedition for its return voyage to India, probably with the hope of securing Parthian aid against Rome. To avoid the Romans, the expedition sailed southward from Britain and circumnavigated Africa, where monuments were left to mark its course. The date of the expedition's return to the mouth of the Indus has not yet been identified. The earliest possible date would be about A.D. 48. [1, p. 5]

This is all news to historians, but is only a prelude to a central figure in AIM's scheme of things, Apollonious (so spelt by AIM) of Tyana, who

> lived when Rome was ruled by insane tyrants and their secret police, by Caligula, Nero, and Domitian; when true science was mocked as valueless and philosophers were persecuted. He could foresee that European society and science would continue to degenerate under Roman tyranny, and he could reasonably anticipate many centuries of intellectual darkness. His problem was therefore to devise a system of public records, invisible to Romans and other anti-scientific persons, which would outlast the dark age of Roman civilization and disclose itself to the scientists of some more enlightened future age. He solved this problem by applying to the medium of religious architecture the principle of the mathematical puzzle. By inventing puzzles that only mathematicians could recognize and read, he assured that the meaning of his record would be conserved for a society which valued mathematical science. He and his school invented a vast complex of interlocking puzzles based primarily on time and distance measure numbers: the dates of mathematicians, the latitudes and longitudes of cities associated with them, the dimensions of buildings in these cities, the boundaries of continents and islands, the location of principal rivers, lakes, and mountains, the distances and dimensions of the Sun and planets, and many other facts—natural, historical, or mythical—which could be expressed in measure numbers. These puzzles were then recorded in the form of monuments, dated and undated mathematical inscriptions, and writings in Pythagorean cipher.
>
> [1, pp. 7–8]

Let us look at an example,

> one of the simplest puzzles recorded in the *Life* of Apollonious [by Philostratus, translated by F. C. Conybeare]. In bk. i ch. vii we are told that Apollonious was guided by his tutor Euxenes until... when he reached his sixteenth year he indulged his impulse towards the life of Pythagoras, being fledged and winged by some higher power. Notwithstanding he did not cease to love Euxenes, nay, he persuaded his father to present him with a villa

> outside the town, where there were tender groves and fountains, and he said to him: "Now you live there your own life, but I will live that of Pythagoras." [1, pp. 12–14]

That statement may not strike you as noteworthy, but let us look at what AIM makes of it. They use gematria, of course. First we need the value of "Pythagoras":

Π	Υ	Θ	Α	Γ	Ο	Ρ	Α	Σ
80	400	9	1	3	70	100	1	200

The total is 864. AIM says that we need not just one but three numbers for each name. Besides the sum of the values of the letters, as above, we need to sum the *pythmenes*, which are the sums of the digits in each letter. That is, we throw away the zeros. The sum of Pythagoras's pythmenes is

$$8 + 4 + 9 + 1 + 3 + 7 + 1 + 1 + 2 = 36.$$

Finally, we reduce that sum to a single digit, obtaining the *pythmen* for Pythagoras of $3 + 6 = 9$.

A person's *long name* has the value determined by multiplying the first and last of the numbers: for Pythagoras, it is $864 \cdot 9 = 7776$. The corresponding numbers for Apollonius are

Α	Π	Ο	Λ	Λ	Ω	Ν	Ι	Ο	Σ
1	80	70	30	30	800	50	10	70	200
1	8	7	3	3	8	5	1	7	2

Name, 1341; sum of pythmenes, 45; pythmen, 9; long name, 12069. Similarly for Euxenes:

Ε	Υ	Ξ	Ε	Ν	Ο	Ν
5	400	60	5	50	70	50

yielding 640, 37, and 10. (The reader may wonder why the pythmen is not 1. I wonder, too. There may be a rule that the pythmen, since it is used as a multiplier to get the long name number, must be greater than one. It is also possible that if it were not 10, this puzzle would not work out the way that AIM wanted it to.)

Now we are ready to go. When we take the eleven words in Apollonius's statement and gematrize we get a total, sum of pythmenes, and sum of pythmens of 5901, 231, and 69. Since Euxenes was going away, we clearly need to subtract his numbers from Apollonius's statement:

Statement	5901	231	69
Euxenes	640	37	10
Difference	5261	194	59

Apollonius was now going to act on his own responsibility, "so," says AIM, "we will add the numbers of Apollonious":

Numbers so far	5261	194	59
Apollonius	1341	45	9
Sum	6602	239	68

Now it is time to add the numbers:

$$6602 + 239 + 68 = 6909.$$

Apollonius was going to live the life of Pythagoras, so we add his number:

$$6909 + 864 = 7773.$$

Since there are three people involved, Euxenes, Apollonius, and Pythagoras, "we will therefore add 3 to the total":

$$7773 + 3 = 7776.$$

This is the long name of Pythagoras!

AIM says that

> the puzzle of the sixteenth year of Apollonious verifies, by one simple example, our hypothesis that the sayings and writings of Apollonious are expressed in the pythagorean calculus.

Those who are inclined to cavil at numerology could point to the calculation that gave the total of 7773 and ask, why add only the value of the name of Pythagoras? Euxenes and Apollonius carry three numbers apiece; why only one for Pythagoras? Cavilers think that they know the answer, namely that without such fiddling it would be impossible to reach the final number.

If you do not cavil, then think of what Apollonius had to go through to construct his puzzle! He had to think of a statement, appropriate to the occasion of dismissing Euxenes, that would have *precisely* a sum of 5901, a sum of pythmens of 231, and a sum of pythmenes of 69. He not only did it, if AIM is correct, he did it in such a masterly fashion that his remark sounded perfectly natural, as if he tossed it off without thinking. This is not an easy feat. Try to construct a sentence that would gematrize to 5901 (or 6000, or any preset goal) expressing something simple, such as "Have a nice day" and see what luck you have. Of course, AIM could reply that the difficulty we have merely shows how far short we fall of Apollonius in intellectual and numerological capacity.

Richard Guy, the enunciator of the Law of Small Numbers (chapter 9), notes that if we assign numbers to letters with $a = 0, b = 1, c = 2, \ldots, z = 25$, then the numerical values of the words in "have a nice day" are

have	32
a	0
nice	27
day	27

The product of the nonzero numbers is 46656 which is $6 \cdot 7776$, precisely six times the long name of Pythagoras! The significance of this (and significance it must have, since it could not be conicidence, could it?) remains to be worked out.

The reason that Apollonius's statement has eleven words, by the way, is that he was sixteen years old. Here is how AIM gets that:

> If we read this to mean 16 years of 365.2422 days each, we find the numerical coincidence between the period of time and the numerical value of Apollonious' statement:
> (1) 16 years $\times$ 365.2422 days = 5843.8752 days
> (2) 5901 (the value of A's statement by gematria) $-$ 5843 = 58
> (3) 69 (the pythmen of A's statement) $-$ 58 = 11, the number of words in A's statement. [1, p. 13]

Apollonius no more knew the length of the year to seven significant digits than Pythagoreans constructed telescopes centuries before Galileo, but that does not matter to AIM. When numbers do such wonderful things, dull facts must take second place.

And numbers can do wonderful things.

> If it be argued that Apollonious reached his "sixteenth year" on his *fifteenth* birthday, so that the quantitative reference is to 15 $\times$ 365.2422 days the dispute can be settled amicably in a Pythagorean manner by operating with the neglected fraction 0.8752, in step (1) as follows:
> (1) 5843 days $\times$ 0.8752 = 5113.7936 days which is a close approximation to 14 years $\times$ 365.2422 days = 5113.3888 days.
> (2) The mean of 16 years and 14 years is 15 years. [1, p. 13]

Let us consider a few more examples of the AIM's numerology. The point of doing so is not solely to laugh at their absurdity. It is also to show what can be done with numbers. *Anything* can be done with numbers. Whatever it is that you want people to believe—that Stonehenge was built by one of the ten lost tribes of Israel; that survivors of the drowning of Atlantis designed the Great Pyramid; that the end of the world will arrive on August 20, 1953; that ancient Pythagoreans carved the statues on Easter Island—you can, given sufficient ingenuity, "prove" it with numbers. A secondary point is to show that numerologists are not all ignorant and only half literate. Many are, but the leader of AIM was highly educated and wrote lovely prose. It goes to show that education alone will not solve all of society's problems.

Here is another example of Apollonius's ability to construct statements with numerological significance. What is remarkable about the statement in the next example is that, seemingly, he *could not have prepared it beforehand.* How could that be?

> Apollonious was brought before a satrap who asked, "... who he was to come trespassing like that into the king's country, and Apollonious said, 'All the earth is mine, and I have a right to go all over it and through it.' " We are required to verify, by mathematical reasoning, that Apollonious did in fact have a right to go over and through the earth.
>
> The Greek text of A's statement contains 10 words which yield by gematria the identifying number 2685...
>
> We elect to solve this puzzle by operating with two given numbers: 10, the number of words in A's statement, and 2685, the numerical value of these words by gematria. It should be noted that 10 is the sacred Pythagorean number, the *tetraktys*, which is associated by tradition with Croton, Italy where it was used as a "password." The steps of our solution are:
>
> (1) 2,685 $\times$ 10 = 26,850.
>
> (2) A's statement refers to the "whole earth" and his right to "go all over it and through it." By the convention of Hipparchus (fl. 161–126 B.C.) a great circle of the earth measures 360° or 21,600′ and the earth's diameter is 6876′. Thus, if A could exercise his right to go over and through the earth, the ideal distance he would travel would be: 21,600′ + 6876′ = 28,476′.

Is that not clever? *Through* the earth means *through*, from pole to pole, or from any point straight down to its antipodal point. The circumference of a circle is π times its diameter, so the number of minutes in a diameter (no one, not even AIM, measures straight lines in minutes of arc, but that is a mere cavil) is thus $21600/\pi = 6875.4935\ldots$, which does not quite round to 6876. But we will let that pass.

> (3) The difference between steps (1) and (2) is: 28,476′ − 26,850′ = 1,626′.
>
> (4) To identify the Pythagorean meaning of this remaining distance with ceremonial precision, we will add to it the value of our operator, 10, the tetrakys: 1,626′ + 10′ = 1,636′ = 27°16′.
>
> (5) Apollonious made his statement at Babylon which is at Long. E. 44°24′. He reached Babylon by travelling east. To identify his ideal point of departure we will measure 27°16′ *west* of Babylon: Babylon at Long. E. 44°24′ − 27°16′ = Long. E. 17°08′ which is the longitude, to the nearest minute, of Croton, Italy, where Pythagoras founded pure geometry (which means literally "earth measurement") and invented geometrical proof!
>
> This puzzle tells us that Apollonious was a Pythagorean philosopher who, by his mastery of the problem of measuring longitude, did in fact have the right and power to go all over the earth. His claim to being a true

> Pythagorean is validated by his accurate measurement of the longitudes of Croton and Babylon. [1, pp. 14–15]

Though there is a wealth of examples in AIM's 40-page booklet, let us consider only one more, a numerological proof that the statues of Easter Island were carved by Pythagoreans as part of their voyage around the world. I reluctantly omit the numerological demonstration that the Mayan civilization was founded by Pythagoreans, and that the Maoris of New Zealand are descended from Indians of the Maurya dynasty.

Easter Island, "one of the most isolated islands in the world... about 1,400 miles from Pitcairn, the nearest island to the west, and about 2,000 miles west of South America," [1, p. 19] is known for its huge statues, all variations on a theme: all of men, distorted so that 40% of their bodies are taken up with their heads, with long brooding faces and large ears, of heights mostly between 10 and 15 feet, but ranging up to 35 feet. There are almost 600 of them. They bear some similarities to statues found in the Andean highlands of South America, though the island was probably settled from the west, some two thousand years ago. The statue-carvers left some examples of an ancient script, but these have not been deciphered. By the time Europeans came upon the island in 1722, all memory of the how and why of the statues had departed from the inhabitants. So, it is legitimate to state that the identity of the creators of the statues and the purpose for which they were made are both mysteries.

But not to AIM. Numbers dissolve mysteries. It is important to keep in mind that

> a number of the statues were mounted on stone platforms near the shore. They were crowned with cylindrical hats carved from a rose-colored volcanic tuff. [1, p. 20]

Also to be kept in mind is that one of the seven wonders of the ancient world was the Colossus of Rhodes, a large statue (some have speculated that it was so large that ships sailed between its legs at the entrance to the harbor at Rhodes).

Here we go. Hold on to your hats, cylindrical or otherwise. The numbers may make you dizzy.

> (1) We identify the statues as conventional Greek boundary markers erected for the guidance of travellers and known as Hermae.
>
> (2) The Greek name for this type of statue is the name of the god Hermes:
>
> | E | P | M | H | Σ | |
> |---|---|---|---|---|---|
> | 5 | 100 | 40 | 8 | 200 | = 353 |
> | 5 | 1 | 4 | 8 | 2 | = 20 |
>
> Pythmen: $2 + 0 = 2$.

(3) We identify the cylindrical hats as a circular head covering or cap, in Greek:

Π	Ε	Ρ	Ι	Σ	Τ	Ε	Γ	Ω
80	5	100	10	200	300	5	3	800
8	5	1	1	2	3	5	3	8

1503, Pythmen: $3 + 6 = 9$.

(4) When we subtract the pythmen of the cap from its numerical value by gematria, $1503 - 9 = 1494$, and subtract the numerical value of the statue from the cap, $1494 - 353 = 1141$, the result is the identifying number of Hipparchus:

Ι	Π	Π	Α	Ρ	Χ	Ο	Σ
10	80	80	1	100	600	70	200
1	8	8	1	1	6	7	2

1141, Pythmen $3 + 4 = 7$.

(5) The hats are rose-colored and, by direct association in this context, suggest the name of the city of Rhodes which is derived from the Greek word for rose.

(6) Hipparchus founded the science of measuring longitude and established the world's first prime meridian at the Port of Rhodes at Lat. N. 36°27′ and Long. E. 28°14′. To find the longitude of Easter Island from the Port of Rhodes we operate with the three identifying numbers of Hermes—(the god of travel and the name for each statue)—as follows: Measure $(353\times20\times2)-[(353\times2)+(20\times2)+(20+2)] = 13{,}352$ minutes of longitude or 222°32′ east of the Port of Rhodes at Long. E. 28°14′. The result is Long. W. 109°14′ or one minute east of Easter Island!

(7) This solution is based upon the pythagorean calculus invented by Pythagoras of Samos. The city of Samos is at Lat. N. 37°41′. To reach the latitude of Easter Island from Samos we operate with the numbers of Hermes as follows: Measure $353+(353\times20/2) = 3883$ minutes of latitude or 64°43′ of latitude south of Samos, i.e., 64°43′ − N. 37°41′ = S. 27°02′, or one minute north of Easter Island!

(8) Thus we see that the colossal Herms of Easter Island do in fact inform the Pythagorean traveller about the latitude and longitude of this island. [1, pp. 20–21]

If you were a Pythagorean traveler, washed up on the shore of Easter Island, would those calculations occur to you? It is conceivable that a Pythagorean traveler could think, "Why those statues, they must be herms! Hermes, let me see... yes, 353, with sum of pythmenes 20 and pythmen 2." Pythagoreans would be trained to think that way. But how would you know to combine the 353s, 20s, and 2s in the very different ways of (6) and (7) to determine your longitude and latitude?

You wouldn't know. The calculations are absurd, would occur to no traveler, Pythagorean or otherwise, and were done for the first time by AIM. The reason that I am presenting so much absurdity in such detail is to demonstrate that a large number of absurdities do not combine to create something that may be true. This is obvious, but it is overlooked by people who say, about pyramidology, Stonehengery, or some similar numerology, "But there is *so much* here! It can't be by accident! There *has* to be something behind it." No, there does not have to be anything behind it. Nonsense piled upon nonsense does not make sense.

You might wonder, why choose Easter Island to build statues on? It was no accident. It was the golden mean, $\varphi = 1.618\ldots$ (see chapter 29) that made the Pythagoreans do it. First, we note that 40% of the Easter Island statues were taken up with their heads, leaving 60% for their bodies. That is close enough to $1/\varphi = .618\ldots$ to show that the statue-carvers had φ in mind. The Colossus of Rhodes was, according to AIM, 105 feet high. Sources I looked at put its height at something around 120 feet, but 105 feet is close enough. The largest Easter Island statue, according to AIM, is 65 feet high, and $65/105 = .619\ldots$ is the golden mean, almost exactly. I do not know where AIM got that 65 feet, since most authorities put the height of the largest statue at under 40 feet. Could it be that AIM is following the numerological principle that if your numbers do not give you what you want, you may make some new ones up? But we must be fair to the AIM: it is possible that someone once found an Easter Island statue that looked as if it was, or had been, 65 feet high.

Be that as it may, AIM goes on:

> Rhodes is historically associated with the science of longitude. We will therefore measure the difference of longitude between the Colossus of Rhodes and the colossus of Easter. The former is at Long. E. $28°14'$ and the latter at Long. W. $109°17'$. The difference of longitude is therefore $137°31'$ which is, to one minute, the lesser segment of a great circle cut in extreme and mean ratio. [1, p. 22]

There follows a picture like Figure 1, showing that from Rhodes to Easter Island is $1/\varphi$ of a complete circle of longitude.

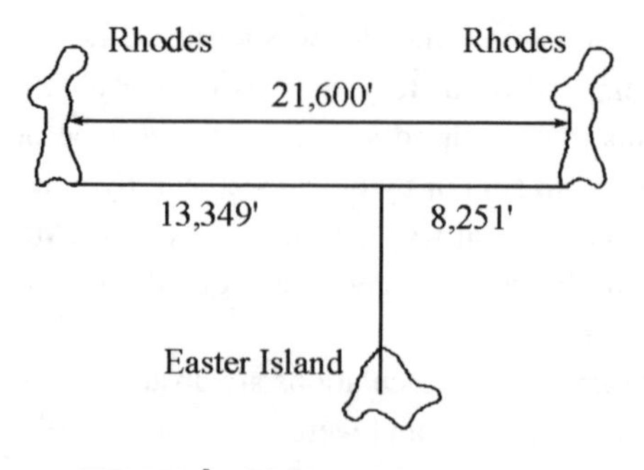

FIGURE 1 Rhodes to Easter Island.

This is remarkable, but presenting one remarkable coincidence does not reveal on which try the coincidence occurred. AIM may have looked first at changes in latitude to see if any φ could be teased out of them. They may have started at Samos (for Pythagoras) instead of Rhodes (for Hipparchus) only to find no φs. There are many things they could have tried, but the failures are not reported. This also should be kept in mind when examining the startling findings of numerologists.

We will now give AIM a rest. If you want the proof that the Orestes plays of Aeschylus were performed in the ancient Central American city of Palenque, you will have to find your own copy of [1]. The founder of AIM was a person of high but misguided ingenuity, whose organization probably died with him. Unfortunately, he has had, and continues to have, successors.

Reference

1. *Special Publication 1*, American Institute of Man, Chicago, 1950.

CHAPTER 5

Alphabets for Gematria

Gematria is the ancient art and science of turning words and names into numbers. The reason for doing this is to find out more about the word, the name, or the object or person that they describe. The findings are, of course, mystical.

Gematria goes back a long way. Its first appearance in literature was c. 200 A.D., though it is possible that its practice goes back much further. For example, in Genesis 14:14, we are told that Abram took his servants, three hundred and eighteen of them, off to rescue his brother and then in 15:2 learn that the name of his steward was Eliezer. The value of "Eliezer" is 318. This may be coincidence, but more likely is gematria.

The mystical basis of gematria is given in *The Jewish Encyclopedia*:

> All creation has developed through emanation from the En Sof. The first degrees of that emanation are the ten SEFIROT, from the last of which, Kingdom, developed the twenty-two letters of the Hebrew alphabet. Through the latter the whole finite world has come into existence. Those letters are dynamic powers. Since these powers are numbers, everything that has sprung from them is number. Number is the essence of things, whose local and temporal relations ultimately depend on numerical proportions. Everything has its prototype in the world of spirit, that spiritual prototype being the germ from which the thing has been developed. As the essence of things is number, the identity of things in number demonstrates their identity in essence.
>
> [3, vol. 6, p. 590]

"The essence of things is number": was Pythagoras influenced by the ancient Hebrews, or was it the other way around?

To practice gematria, you need to be able to convert words into numbers. If you asked the average person how to do this, the average answer I think would be

A	B	C	D	E	F	G	H	I
1	2	3	4	5	6	7	8	9
J	K	L	M	N	O	P	Q	R
10	11	12	13	14	15	16	17	18
S	T	U	V	W	X	Y	Z	
19	20	21	22	23	24	25	26	

That is the alphabet that is most often found among general-purpose gematrists. But if you are looking to put 666, the number of the beast (considered in more detail in the next two chapters), on someone or something, this natural alphabet will not work for short words. The average letter has value 13, so a total of 666 will take, on the average, around fifty letters. A way around this is to add 100 to each value. Six letters will then have an average sum of 678, close enough to 666 to guarantee some hits. A variation of this, gotten by adding 99 instead of 100, was used to make "Hitler" a beast:

A	B	C	D	E	F	G	H	I
100	101	102	103	104	105	106	107	108
J	K	L	M	N	O	P	Q	R
109	110	111	112	113	114	115	116	117
S	T	U	V	W	X	Y	Z	
118	119	120	121	122	123	124	125	

Sure enough: $107 + 108 + 119 + 111 + 104 + 117 = 666$.

This idea can be used to beast anyone or anything that has six letters that sum, in the ordinary alphabet, to 66 or 66 plus or minus any multiple of 6. That may be the way the beasthood of Hitler was discovered: in the ordinary alphabet, "Hitler" is $8 + 9 + 20 + 12 + 5 + 18 = 72$, just six over 66, so adding six 100s and subtracting six 1s—that is, using the last alphabet—gives the toal 666. Since the chance that a six-letter word will have such a sum is one in six, beasts with six letters are not hard to locate. For example, President Clinton can be beasted: "Clintn" (that's how it's pronounced, after all) has a sum of $3 + 12 + 9 + 14 + 20 + 14 = 72$, so the Hitler alphabet will do it. If the sum were 60, an alphabet starting with $A = 102$ would do the trick. Start with $A = 98$ and "Luther" will gematrize to 666.

To take care of people or things with other than six letters, other shifts can be used. If something has a sum in the ordinary alphabet that differs from 666 by a multiple of the number of letters in its name, then a shift will produce a beast. For example, numerologists could establish, not that they would want to, that numbers themselves are the beast. The value of "numbers" is

$$14 + 21 + 13 + 2 + 5 + 18 + 19 = 92$$

and $666 - 92 = 574 = 82 \times 7$. That is, $92 + 82 \times 7 = 666$, so adding 82 to each of the seven letter values will produce 666. That can be done by using the alphabet that starts with a = 83, b = 84, and ends with z = 108. The only problem would be to find a sound numerological reason why that alphabet is the appropriate one. Its middle value is 96, suggesting 1996, but why numbers would be beastly only in that year is not clear either.

A final variation on the shift idea is the shift with reduction modulo 26. For example, an alphabet shifted over by seven (and a good numerological reason could no doubt be found for shifting that mystical number of units) and reduced would be

A	B	C	D	E	F	G	H	I
8	9	10	11	12	13	14	15	16
J	K	L	M	N	O	P	Q	R
17	18	19	20	21	22	23	24	25
S	T	U	V	W	X	Y	Z	
26	1	2	3	4	5	6	7	

Combining that with previous ideas enlarges still further the scope and range of beasting.

These alphabets, natural as they are, are not in keeping with the historical development of gematria, which used the Greek or Hebrew alphabets, where letters *were* numbers. The ancient Greeks, who first had the idea, used the correspondence we have seen in chapter 3:

α	β	γ	δ	ε	ϛ	ζ	η	θ
1	2	3	4	5	6	7	8	9
ι	κ	λ	μ	ν	ξ	ο	π	ϙ
10	20	30	40	50	60	70	80	90
ρ	σ	τ	υ	φ	χ	ψ	ω	ϡ
100	200	300	400	500	600	700	800	900

The advantage of this system of writing numbers, as compared with those that had gone before (and some that came after), was that it represented numbers compactly. If the ancient Egyptians or Babylonians wanted to write a 4, they had to put down four copies of the symbol for the unit. A 44 would take eight symbols, instead of a quick μδ. Even the Roman XLIV is more work to write and is slightly harder to read.

A disadvantage of the Greek system is that it would be harder to learn, at first, than other systems. It might be thought that it would be more difficult to calculate with—think of having to memorize all possible products of two Greek letters—but it was not. Ancient peoples carried out their calculations on an abacus, or by sliding tokens along lines, or moving pebbles in grooves, and used their number symbols only for recording the results. Roman numerals are as good as any for that purpose, and it was not until the seventeenth century that they started to fade away in Europe. It was only after paper became cheap that it could be used for anything as unimportant as numerical scratch work.

A byproduct of the Greek system was gematria. Every word was also a number, and some numbers were also words. The idea of associating the two was not long in coming to the agile minds of the ancient Greeks. Gematria would not have been unleashed on the world if the Greeks had not had the idea of using the alphabet for numbers. The Greeks got their alphabet from the Phoenicians sometime around the tenth century B.C., but it was not until several centuries later that they began to use it to write numbers. Until the fifth century B.C. they used a system similar to the later Roman numerals, for example writing 2020 as XXΔΔ, two capital chis and two capital deltas for two thousand (chilioi—χιλιοι) and two tens (deka—δεκα). It was not until the first century B.C. that alphabetic numerals were made official in Athens. (This information is taken from that admirable book, *Number Words and Number Symbols* by Karl Menninger [2].)

The Greeks partially repaid their debt to the East by giving to the Hebrews the idea of using letters for numbers, and that is what the Hebrews did:

א	ב	ג	ד	ה	ו	ז	ח	ט
1	2	3	4	5	6	7	8	9
י	כ	ל	מ	נ	ס	ע	פ	צ
10	20	30	40	50	60	70	80	90
ק	ר	ש	ת					
100	200	300	400					

Users of the Hebrew alphabet also took up the sport of gematria, and they continue it to this day. Recently a book was published that gave the numerical value of each word in the Pentateuch, arranged in increased numerical order. An advertisement for it read:

> Finally available! A handy one-volume book containing all the words of the Chumach—in Hebrew and in English—listed according to their numerical order. [1]

It mentioned that

> by means of *gematria*, the Rabbis have found supportive evidence that there are 39 categories of forbidden labor on the Sabbath. [1]

Thus does number mysticism—attaining ineffable truths—slide into numerology—applying them to find lucky days, types of forbidden labor, and so on.

During the 1000 years from 500 to 1500, nonreligious intellectual activity was at a low ebb in Europe, gematria included. With the revival of learning came the revival of gematria. A Roman alphabet analogue of the Greek system appeared in 1583:

A	B	C	D	E	F	G	H	I
1	2	3	4	5	6	7	8	9
K	L	M	N	O	P	Q	R	S
10	20	30	40	50	60	70	80	90
T	U	X	Y	Z				
100	200	300	400	500				

Since letters were no longer used to represent numbers, this was exclusively for gematria.

To adapt this to modern English, identify I and J, and U, V, and W. A 1649 variant puts P = 60, R = 70, S = 80, T = 90, U = 100, and Z = 1000. Another alphabet in 1651 inserts W = 300 and moves each of X, Y, and Z up by 100. Yet another in 1681 keeps all up to and including T the same, but assigns to U, V, . . . , Z the values 1000, 2000, . . . , 5000. Clearly, there was not universal agreement among gematrists about the values of the letters. But this was the spirit of the times: the seventeenth century was also a time of instability in mathematical notation (it took Euler finally to set the standard), and English spelling was still in a state of flux. There is no reason to expect gematria to be any different.

A slightly different idea was used in a 1683 alphabet:

A	B	C	D	E	F	G	H	I
1	2	3	4	5	6	7	8	9
K	L	M	N	O	P	Q	R	S
10	20	30	40	50	60	70	80	90
T	U	W	X	Y	Z			
100	110	120	130	140	150			

This is departing from the original Greek. It is the alphabet, with V = 120 added and W, X, Y, and Z each increased by 10, that Tolstoy used in *War and Peace* to beast Napoleon (see chapter 6).

Yet another variation was called in 1707 *Alphabetum Cabbalisticum Vulgare*, the common cabalistic alphabet,

A	B	C	D	E	F	G	H	I
1	2	3	4	5	6	7	8	9
K	L	M	N	O	P	Q	R	S
10	11	12	40	50	60	70	80	90
T	U	W	X	Y	Z			
100	110	120	130	140	150			

The first appearance of the natural-order alphabet was in 1532, when Michael Stifel used

A	B	C	D	E	F	G	H	I
1	2	3	4	5	6	7	8	9
K	L	M	N	O	P	Q	R	S
10	11	12	13	14	15	16	17	18
T	U	X	Y	Z				
19	20	21	22	23				

Stifel also used the trigonal alphabet, assigning letters the successive triangular numbers:

A	B	C	D	E	F	G	H	I
1	3	6	10	15	21	28	36	45
K	L	M	N	O	P	Q	R	S
55	66	78	91	105	120	136	153	171
T	U	X	Y	Z				
190	210	231	253	276				

There are many other alphabets that have appeared in print [4, pp. 130–138], but it is not clear how many were actually used. It is possible that they were the ideas of theoretical gematrists, who presented them with no thought of practical application, even as many modern mathematicians do not care how their theorems are used, or if they are used at all. After thinking of triangular numbers it is no great leap to think of square and pentagonal numbers,

A	B	C	D	E	F	G	H	I
1	4	9	16	25	36	49	64	81
K	L	M	N	O	P	Q	R	S
100	121	144	169	196	225	256	289	324
T	U	X	Y	Z				
361	400	441	484	529				

and

A	B	C	D	E	F	G	H	I
1	5	12	22	35	51	70	98	117
K	L	M	N	O	P	Q	R	S
145	176	210	247	287	330	376	425	477
T	U	W	X	Y	Z			
532	590	651	715	782	852			

One can continue to hexagonal, heptagonal.... Tatlow [4] cites sources for alphabets up to 15-gonal, where Z = 3612. That must have been of theoretical interest only; however, its existence shows that it was of interest. The seventeenth century was an age more golden for gematria than the present.

Going to three dimensions, there are records of pyramidal alphabets, both three- and four-sided. The triangular pyramid numbers are

A	B	C	D	E	F	G	H	I
1	4	10	20	35	56	84	120	165
K	L	M	N	O	P	Q	R	S
220	286	364	455	560	680	816	969	1140
T	U	W	X	Y	Z			
1330	1540	1771	2024	2300	2600			

(There are good numbers there: 220, the smallest member of an amicable pair; 364, almost the number of days in a year; 969, Methuselah's age at death.)

The square pyramid alphabet (see Figure 1), not so far noticed by pyramidologists that I know of, is

A	B	C	D	E	F	G	H	I
1	5	14	30	55	91	140	204	285
K	L	M	N	O	P	Q	R	S
385	506	650	819	1015	1240	1496	1785	2109
T	U	W	X	Y	Z			
2470	2870	3311	3795	4324	4900			

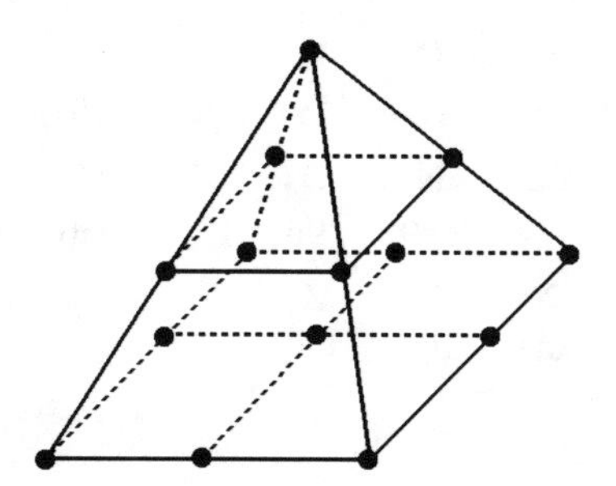

FIGURE 1 A square triangular number.

The alphabets go on and on, in amazing variety. One from 1630 is

A	B	C	D	E	F	G	H	I
1	6	12	20	30	42	56	72	90
K	L	M	N	O	P	Q	R	S
110	132	156	182	210	240	272	306	342
T	U	W	X	Y	Z			
380	420	462	506	552	600			

The numbers are Pythagorean, the pronic numbers: $2 \cdot 3 = 6$, $3 \cdot 4 = 12$, and so on to $24 \cdot 25 = 600$. The first should be $1 \cdot 2 = 2$ to be consistent, but the feeling that A should be 1 seems to have won out.

It won out as well in a 1651 alphabet,

A	B	C	D	E	F	G	H	I
1	6	12	18	24	30	36	42	48
K	L	M	N	O	P	Q	R	S
54	60	66	72	78	84	90	96	102
T	U	W	X	Y	Z			
108	114	120	126	132	138			

where consistency would have 0 for A.

There is no difficulty in seeing how those alphabets arose, but what of this one, attributed to Cornelius Agrippa (c. 1520)? [4, p. 49]

A	B	C	D	E	F	G	H	I
3	3	24	25	3	3	8	15	15
K	L	M	N	O	P	Q	R	S
15	22	23	15	8	13	22	22	9
T	V	X	Y	Z				
5	5	8	3	3				

Or this one, from Cathanus Magus? [4, p. 50]

A	B	C	D	E	F	G	H	I
1	3	22	24	22	3	7	6	20
K	L	M	N	O	P	Q	R	S
1	10	23	12	8	13	27	13	9
T	V	X	Y	Z				
8	2	6	3	4				

They are curious indeed, and what idea was behind them is not at all clear. Though different, the two have too many similarities to be independent inventions. But what it was that inspired them we will probably never know.

There are so many alphabets here, and so many variations on them possible, that a numerologist should not have any trouble in assigning almost any number to almost any name, word, or phrase. The task cries out to be computerized. The user would enter what is to be gematrized, specify what alphabet or alphabets to use, and the easily-constructed program would return numerical values. An advanced version of the program, available at extra cost, could accept a name and the total desired and then find the alphabet that would produce the total. With some extra effort the program could be extended to transliterate into Greek or Hebrew. It would be a boon to numerologists. Luckily, the market is too small. As yet.

References

1. Locks, Gutman G., *The Spice of Torah—Gematria*, Judaica Press.
2. Menninger, Karl, *Zahlwort und Ziffer*, Vandenhoeck and Ruprecht, Göttingen, 1958, translated into English as *Number Words and Number Symbols* by Paul Broneer, M. I. T. Press, Cambridge, 1969, reprinted by Dover, New York, 1994.
3. Singer, Astuter, editor, *The Jewish Encyclopedia*, Funk and Wagnalls, New York, 1907.
4. Tatlow, Ruth, *Bach and the Riddle of the Number Alphabet*, Cambridge University Press, Cambridge, 1991.

CHAPTER 6
The Beast

All numbers have power, but some have more than others. 666 is a number with great power; in this case, the power to inspire vast amounts of speculation and writing, all of it useless and much of it foolish. The mischief-maker who started it all was the author of the book of Revelation, who included in chapter 13, as verse 18:

> Here is the key; and anyone who has intelligence may work out the number of the beast. The number represents a man's name, and the numerical value of its letters is six hundred and sixty-six. [5]

This was the beast who caused, or will cause, everyone "to be branded with a mark on his right hand or forehead, and no one was allowed to buy or sell unless he bore this beast's mark, either name or number"—chapter 13, verse 17. We will subsequently see the connection of this with the Universal Price Code.

Here, advocates of Biblical inerrancy notwithstanding, we have an example of a flagrant error in the Bible. Many people, and many with intelligence, have tried to work out to whom the number of the beast belongs, but who was meant is still uncertain.

New beasts are being found all the time. *Harper's* magazine reprinted an identification of the beast as William Gates, the president of Microsoft:

> Bill Gates's full name is William Henry Gates III. Nowadays he is known as Bill Gates (III). By converting the letters of his current name to their ASCII values, you get the following:
>
> B I L L G A T E S 3
> 66 + 73 + 76 + 76 + 71 + 65 + 84 + 69 + 83 + 3 = 666. [3, p. 24]

This is the first time I know of that the ASCII alphabet, a = 64, b = 65, ..., z = 89, has been used for gematria. The sum is a slight cheat since the ASCII code for "3" is not 3, it is 51, but no alphabet could be more appropriate for a computer person. It may be significant that when 1 is added to the ASCII rendering of "Windows 95", the result is again 666. The unknown (to me) originators of those beastings were just kidding, I think, so we do not have to go into the question of how the writer of Revelation knew about ASCII code, but other beasters have been quite serious.

Irenaeus of Lyon (c. 150–c. 200) is the first writer that we know of who considered the identity of the beast. According to the Seventh-Day Adventist *Encyclopedia* (pp. 898–902),

> he confesses doubt as to the meaning of the number, which suggests that no tradition concerning this had been handed down through the churches of Asia, from which Irenaeus had come, for elsewhere he is in possession of a number of such traditions attributed to John, preserved there.

The best that Irenaeus could do was to find it in "Evanthas,"

Ε	υ	α	ν	θ	α	ς
5	400	1	50	9	1	200

similar (the second α should be an ϵ) to a name given to the pagan god Dionysus. This is hardly convincing—*anyone* who has intelligence can see that?—so he had another shot with "lateinos," Latins:

λ	α	τ	ε	ι	ν	ο	ς
30	1	300	5	10	50	70	200

since "the Latins are they who at present bear rule" and "Titan,"

Τ	ε	ι	τ	α	ν
300	5	10	300	1	50

a variant spelling of Titan, referring to the Titans of Greek myth or perhaps to the Roman emperor Titus.

The preferred assignation of beasthood, taking into account the time when Revelation was written and the obviousness at that time of the beast's identity, has been to Nero. But there are difficulties with this identification. For one thing, it does not work in Greek

Ν	ε	ρ	ο	ν
50	5	100	70	100

but if we take "Neron Kaisar" in Greek and transliterate it into Hebrew, we have

נ	ר	ו	נ	ק	ס	ר
50	200	6	50	100	60	200

just what we need.

This has the advantage of being a possible solution to the mystery of why some ancient texts of the Bible have 616 in place of 666 in the Revelation verse. If we take the Latin "Nero Caesar" and transliterate, we get the letters as above but with a נ missing, giving a total of fifty less, or 616. Some other ancient manuscripts, showing what I think is the great good sense of their copyists, leave the verse out altogether.

Assigning 666 to Nero might be thought to be satisfactory since he was certainly beastly, but it is not. One is that the spelling "qsr" for "caesar" is not quite right, "qysr" being better. This is not too serious since, as we will see, distortions as bad or worse have been made by others to make names come out to a total of 666. Anyway, everyone would know who *Neron Qsr* was. More serious is the objection that the Christian readers of Revelation could not be expected automatically to think of Hebrew gematria, or think of Hebrew at all, on reading the verse. (In verse 16 of chapter 16, "So they assembled the kings at the place called in Hebrew Armageddon," readers were alerted that something Hebrew was coming up.)

Satisfactory or not, the explanation did not keep other people from offering other interpretations. The Venerable Bede (early eighth century A.D.) put forward "antemos,"

α	ν	τ	ε	μ	ο	ς
1	50	300	5	40	70	200

"contrary to honor" and "arnoume,"

α	ρ	ν	ο	υ	μ	ε
1	100	50	70	400	40	5

roughly, "we deny." These were sufficiently farfetched that they attracted no disciples.

Peter Bungus (d. 1601) published *Numerorum Mysteria* (The Mystery of Numbers) in Bergamo around 1584. Bungus goes through the integers 1, 2, 3, . . . , and mentions whatever he could find remarkable connected with each. De Morgan says that he quotes or uses 402 authors [1, vol. 1, p. 55], solid research which could explain why the book went through eight editions by 1617.

De Morgan has this to say about it:

> Bungus goes through 1, 2, 3, etc., and gives account of everything remarkable in which each number occurs; his accounts not always being mysterious. The numbers which have nothing to say for themselves are omitted: thus there is a gap between 50 and 60. In treating 666, Bungus, a good Catholic, could not compliment the Pope with it, but he fixes it on Martin Luther with a little

forcing. If from A to I represent 1–10, from K to S 10–90, and from T to Z 100–500, we see:

M	A	R	T	I	N	L	U	T	E	R	A
30	1	80	100	9	40	20	200	100	5	80	1

which gives 666. Again, in Hebrew, *Lulter* does the same:

ל ו ל ת ר
200 400 30 6 30

And thus two can play at any game. The second is better than the first: to Latinize the surname and not the Christian name is very unscholarlike. The last number mentioned is a thousand millions; all greater numbers are dismissed in half a page. Then follows an accurate distinction between *number* and *multitude*—a thing much wanted both in arithmetic and logic.

What may be the use of such a book as this? The last occasion on which it was used was the following. Fifteen or sixteen years ago the Royal Society determined to restrict the number of yearly admissions to fifteen men of science.... When the alteration was discussed by the Council, my friend the late Mr. Galloway, then one of the body, opposed it strongly, and inquired particularly into the reason why *fifteen*, of all numbers, was the one to be selected. Was it because fifteen is seven and eight, typifying the Old Testament Sabbath and the New Testament day of resurrection following? Was it because Paul strove fifteen days against Peter, proving that he was a doctor both of the Old and New Testaments? Was it because the prophet Hosea bought a lady for fifteen pieces of silver? Was it because, according to Micah, seven shepherds and eight chiefs should waste the Assyrians? Was it because Ecclesiastes commands equal reverence to be given to both Testaments—such was the interpretation—in the words "Give a portion to seven, and also to eight"? Was it because the waters of the Deluge rose fifteen cubits above the mountains?—or because they lasted fifteen decades of days? Was it because Ezekiel's temple had fifteen steps? Was it because Jacob's ladder was supposed to have had fifteen steps? Was it because fifteen years were added to the life of Hezekiah? Was it because the feast of unleavened bread was on the fifteenth day of the month? Was it because the scene of the ascension was fifteen stadia from Jerusalem? Was it because the stone-masons and porters employed in Solomon's temple amounted to fifteen myriads? etc. The Council were amused and astounded by this volley of fifteens which fired at them; they knowing nothing of Bungus, of which Mr. Galloway—who did not, as the French say, indicate his sources—possessed the copy now before me. In giving this anecdote I give a specimen of the book, which is exceedingly rare. Should another edition ever appear, which is not very probable, he would be but a bungling Bungus who should forget the *fifteen* of the Royal Society. [1,vol. 1, pp. 55–57]

A modern successor to Bungus, even to the title, is *The Mysteries of Number* [6]. As the fifteen example above, it also gives various occurrences of small integers. Both books are essentially verifications of the Law of Small Numbers (chapter 9).

Bungus carried on the tradition of ignoring the plain words of Revelation that anyone with intelligence, presumably anyone reading it not too many decades after its composition, would know who the beast was. If it was going to be necessary to wait for one or two thousand years before it became plain, should that not have been mentioned?

With a little straining, it is possible to beast almost anyone. For example, I took care of Ronald Reagan, back when he was a large figure on the world stage, with the $a = 101, b = 102, \ldots, z = 126$ alphabet:

R	E	A	G	U	N
118	105	101	107	121	114

The straining is really minimal since, unlike "Lutera," "Reagun" is what the name actually sounds like and, were it not for tradition, how it could be spelled. Other workers in the field have noticed

Ronald	Wilson	Reagan
6	6	6

Using the same alphabet, as is only appropriate for someone who was Reagan's vice president, we get a total of 666 for

I,	G.	B	U	S	H
109	107	102	121	119	108

Not to be partisan, Michael Dukakis, Bush's opponent in the 1988 presidential election, when referring to himself, also sums to 666 using the Greek alphabet, again only appropriate:

I,	Γ.	Δ	υ	κ	α	κ	η	ς
10	3	4	400	20	1	20	8	100

(The gamma is for "governor," his office when running for president.) Strained, but not much.

Left to the reader are the exercises of beasting more recent politicians, or other people. For example, Michael Steuben notes that in the $a = 100$, $b = 101, \ldots, z = 125$ alphabet, "Holmes" has value 666. Confirmation that Sherlock was indeed the beast is given by his address, 221B Baker Street: 221B must have been preceded by 221 and 221A and, removing the letters, we have

$$221 + 222 + 223 = 666.$$

Sir Arthur Conan Doyle was even more clever than people give him credit for.

Martin Gardner found that in the a = 25, b = 24, . . . , z = 0 alphabet, "Falwell" totals 111. So, if he labors for six days, he is a beast every week. Using the a = 0, b = 1, . . . , z = 25 alphabet, Pat Robertson's television show, with a dash interpreted as a minus sign, is 700 − Club = 666, the beast. That is all in fun, but there are people who take 666 very seriously.

It is also possible to beast groups. A favorite target was the Roman Catholic Church. A more unusual one is women, but that is the one at which William Kendrick Hale aimed. In [2], he noted that "woman" (66) is the sum of "man" (28) and "death" (38). He further exhibited the sum

66	woman
43	wife
70	viper
79	mother
84	daughter
90	sister
234	reproductive in-laws
666	

"No woman shall be saved" is one of his assertions, as is "All women shall be cast out forever into the lake of fire." Mr. H. contained within himself quite a bit of misogyny.

And something else as well. Mr. H.'s place of birth gematrizes as

21	21
88	Botolph
87	Street
80	Atlantic
276	

Then we have, he says, "276 × 2 = 552, the two miles, Luke 21: 1–8," and

180	William Kendrick Hale
184	Jesus the Christ
188	Jesus of Nazareth
552	

Whew! Do we detect megalomania? I am glad that I never ran into Mr. H.

Lest the impression be left that numerology is anti-female, consider the following sum that *I* made up:

28	man
69	husband
66	brute
58	father
48	son
86	brother
149	sex-mad rapist
162	a disgusting pig
666	

My sum, I think, is much superior to Mr. H.'s sum for women and makes it look feeble. However, the only conclusion that anyone is allowed to draw is that *anything* can be made into a beast.

Here is how a character in Tolstoy's *War and Peace* made *himself* into a beast. In Book 9 of that long novel we read that, using the unusual alphabet with "a" to "i" having the values 1 to 9, by ones, and "k" to "z" the values 10 to 160, by tens,

> writing the words *L'Empereur Napoléon* in numbers, it appears that the sum of them is 666,

[including a 5 for the "e" dropped in the elision *L'*]

> and that Napoleon was therefore the beast foretold in the Apocalypse. Moreover, by applying the same system to the words *quarante-deux* [forty-two], which was the term allowed to the beast that "spoke great things and blasphemies," the same number 666 was obtained, from which it followed that the limit fixed for Napoleon's power had come in 1812 when the French emperor was forty-two. The prophesy pleased Pierre very much and he often asked himself what would put an end to the power of the beast, that is, of Napoleon, and tried by the same system of using letters as numbers and adding them up, to find an answer to the question that engrossed him. He wrote the words *L'Empereur Alexandre, La nation russe* and added up their numbers, but the sums were either more or less than 666. Once when making such calculations he wrote down his own name in French, Comte Pierre Besouhoff, but the sum of the numbers did not come right. Then he changed the spelling, substituting a *z* for the *s* and adding *de* and the article *le*, still without obtaining the desired result. Then it occurred to him: if the answer to the question were contained in his name, his nationality would also be given in the answer. So he wrote *Le russe Besuhof* and adding up the numbers got 671. This was only five too much, and five was represented by *e*, the very letter elided from the article *le* before the word *Empereur*. By omitting the *e*, though incorrectly, Pierre got the answer he sought, *L'russe Besuhof* made 666. This discovery excited him. How, or by what means, he was connected

> with the great event foretold in the Apocalypse he did not know, but he did not doubt the connection for a moment.

Such is the power of numbers.

John Taylor, the author of *The Great Pyramid. Why was it built? And who built it?* (1859), one of the cornerstones of pyramidology, had a shot at the beast in *Wealth the Name and Number of the Beast* (London 1844) (quoted in De Morgan [1, vol. 1, p. 352]), finding it in

ε	υ	π	ο	ρ	ι	α
5	400	80	70	100	10	1

a word, De Morgan said, used in the Acts of the Apostles for "wealth in one of its most disgusting forms." He added that "this explanation has as good a chance as any other," which is precisely true.

To show how the art of beasting, as any other art, progresses, here is an extension of Taylor's work. In Woodrow's *Babylon Mystery Religion*, we find that

> in the New Testament, the letters of the Greek word "euporia," from which the word WEALTH is translated, total 666. Out of all the 2,000 Greek nouns of the New Testament, there is only one other word that has this numerical value and that word is "paradosis," translated TRADITION. (See Acts 12:25; Mt. 15:2). Wealth and tradition—strangely enough—were the two great corruptors of the Roman Church! Wealth corrupted in practice and honesty, and tradition corrupted in doctrine.

When you combine this with other indications, such as "lateinos" gematrizing to 666 in Greek, "Neron Caesar" in Hebrew, "Filii Vicarius Dei" in Roman numerals—

F	I	L	I	I	V	I	C	A	R	I	U	S	D	E	I
	1	50	1	1	5	1	100			1	5		500		1

and "Romulus," in its Hebrew form, "Romith" likewise, you have no trouble concluding who the beast is. It has to be Rome.

But there have been other opinions. David Thom (quoted in [1, vol. 2, pp. 227ff]) compiled in *The Number and Names of the Apocalyptic Beasts* (London, 1848) all that he could find. Among those in Greek—translated here—were, from Irenaeus down,

Latin	the Latin kingdom	the Italian church
blooming	Titan	renounce
the lustrous	conqueror	bad guide
truthful harmful one	slanderer of old	unmanageable lamb

Antemos	Genseric	Benedict
Boniface III	baneful	the papal brief
Lutheran	Saxon	Beza antigod
Mahomet	Iapetos	Paspeisoks
Diocletian	Braschi	Bonneparte
the megatherium		

Those are only the ones in Greek, and only some of them. In Latin, it is possible to beast

Julius Caesar	Diocletian	Louis XIV
Gerbert	Linus	Pope Paul V
Luther	Calvin	Laud

and so on. Then there are the Hebrew beasts, who also exist in abundance and are by no means all Roman.

Evidently, almost anyone can be made into a beast, so very little weight should be given to any fresh specimen. With the wide variety of alphabets available, using a little ingenuity should make it possible to beast anyone you do not like.

As the new millennium approaches and end-of-the-worlders proliferate we can expect to see ever more beasting. Bar codes, those laser-readable symbols that now appear on almost everything that can be bought, have been seized upon as evidence of the imminent fulfillment of apocalyptic prophesies. A newspaper advertisement, inserted by the Full Gospel Illinois Church of Chicago, points this out. I have no citation for the source, but it can be dated as before October 1992, because that is when the church looked for the onset of the Rapture. The church also warned us to look out for bar codes:

> The familiar barcode is the 666 mark about which the Bible speaks. Every barcode is framed in three 6's. The US and European governments are planning on putting these marks on people's foreheads or right hands by method of laser. This mark will replace cash, credit cards, and identification cards. People without knowledge will welcome this mark for its convenience and privileges. However, we beg you, DO NOT RECEIVE THE 666 MARK! It is better to refuse the mark and be killed than to suffer the eternal consequences for receiving it!

How the church and other similar reasoners got the idea that bar codes are the mark of the beast can be seen in the facsimile of a Universal Product Code (UPC) in Figure 1. The 6 in the right-hand half of the symbol looks as if it is represented by a pair of thin vertical lines. There are also pairs of vertical lines at the beginning, middle, and end of the code—6, 6, and 6. The mark of the beast! Obvious, is it not?

FIGURE 1 A simulated UPC.

No, it is not. Michael Keith explained it all clearly in [4] and I will repeat his explanation here, though not in the hope of changing the minds of the members of the Full Gospel Illinois Church. They probably have minds that cannot be changed, except perhaps about the date of the onset of the Rapture. (The 1992 date was arrived at by taking 1948, the date of the founding of Israel, adding 51 years to take account of "This generation will by no means pass away till all these things are fulfilled. Matthew 24:32, 34" and then subtracting seven for the years of the Great Tribulation. The Tribulation not yet being underway—"There will be wars, world-shaking earthquakes, famines, various plagues, and hail and fire mixed with blood burning a third of the earth"—a changing of minds was necessary.)

The UPC encodes numbers very carefully so that errors will be almost impossible to occur. To code the integers $0, 1, \ldots, 9$, four binary digits would be plenty, but the UPC uses seven, so that each digit differs from each other in at least two places. Further, the digits have different codes in the first and last halves of the symbol. The left-hand digits identify the manufacturer of the product and the right-hand ones identify the product. The binary codes for the left-hand digits are

0	0001101	5	0110001
1	0011001	6	0101111
2	0010011	7	0111011
3	0111101	8	0110111
4	0100011	9	0001011

On the symbol, 1s are thin vertical lines and 0s are blanks, and there is no space between consecutive 1s. Thus a 3 looks like a space, followed by a very thick bar (four lines thick), followed by a thin space, followed by a thin bar. On the right-hand half of the symbol, the digits are coded with the complements of their codes on the left, so that the code for 6 is 1010000. Since the spaces are invisible, a 6 on the right looks like two thin vertical lines separated by a thin space.

UPCs have at their start a 101, whose purpose is to tell the computer that the code is starting and a digit will soon be arriving; at their centers a 01010, saying to the computer that here is the middle; and at their ends another 101, signifying the end. None of these is a 6, but each looks like a six, with the two bars separated by a space. This is how the 666 idea got started. It could look as if UPCs were marked with the number of the beast, which is highly disturbing if you are disturbed by such things.

Thank heavens, not everyone is disturbed by 666. In 1988 the owner or owners of 666 North Lake Shore Drive in Chicago changed the number of the building to 668. The Chicago *Tribune* reported that its tenants protested: they liked the old number [7].

The Tribune also reported in 1994 that the New Mexico Highway Commission declined to change the number of U. S. highway 666 [8]. It was given its number in 1926 following the rule that highways associated with main roads be given numbers of the main road with a digit prefixed, so routes numbered 166, 266, 366, . . . , went along with highway 66. Highway 66 no longer exists, having been obliterated by interstates 44 and 40, so the number no longer makes any logical sense. The government of New Mexico thus deserves all the more credit for resisting number foolishness.

In contrast with those examples of sturdy good sense in the U. S. A. there is this clipping, of whose source I know nothing more than it appeared in an Australian newspaper:

> LONDON: The Devil's number, 666, is proving a problem for British motorists.
>
> "The number of the Beast" has been involved in so many accidents the Driver and Vehicle Licensing Agency has scrapped it.
>
> "People complained of funny things happening whenever they came up against a vehicle with the number," a spokesman for the agency said yesterday.
>
> "There was so much hassle about bumps, collisions and accidents with this number that it was agreed that 666 was not to be used on any vehicle registration."

It is possible that this item appeared in an Australian equivalent of one of our supermarket tabloids so, until its source is found, it is premature to pat ourselves on the back for our superiority over the superstitious British.

The future of 666 is not clear. I think that 666ers take it less seriously than, say, pyramidologists take their pyramid, but see chapter 8 for a serious 666 person. At least, 666ers do not seem to devote their lives to their number—it is just another indication of more important things, the coming of Armageddon and so on. (When that time comes, aramageddon out of here.) But, as we still have buildings that lack thirteenth floors, 666 will be with us for a while, maybe even until the end of the world.

About the end of the world, watch out for September 2, 2037. That date, written as 9/2/37, has a product of $9 \cdot 2 \cdot 37 = 666$. 2037 will be a year in which *five* days, the maximum number possible, have a product of 666, and that day is the last of them. The days are 1/18/37, 2/9/37, 3/6/37, 6/3/37, and 9/2/37. This is a numerological discovery made by me, published here for the first time in a book. Even if none of the days marks the winding up of the world, it would be prudent, on any of them, not to walk under any black cats nor to let any ladders cross your path. By the way, it is not generally known that the lumber of the beast is 5328 ($2 \times 4 \times 666$). But perhaps it is not fitting to joke about such sacred subjects.

The future of 666 is almost as unclear as its past. We will probably never know exactly who the writer of Revelation had in mind. I wish someone could find out.

References

1. De Morgan, Augustus, *A Budget of Paradoxes*, London, 1872, reprinted with notes by D. E. Smith, Open Court, Chicago, 1915, reprinted by Dover, New York, 1954.
2. Hale, William Kendrick, *Rod of Iron*, Hale Research Foundation, Washington, 1954.
3. *Harper's*, February 1995, 24.
4. Keith, Michael, The bar-code beast, *The Skeptical Inquirer*, 12 (1988) #2, 416-418.
5. *New English Bible*, Oxford and Cambridge University Presses, 1961, 1970.
6. Schimmel, Annemarie, *The Mystery of Numbers*, Oxford University Press, New York, 1993.
7. Chicago *Tribune*, April 25, 1988, section 5, pages 1, 2.
8. Chicago *Tribune*, March 6, 1994, section 17, page 6.
9. Woodrow, Ralph, *Babylon Mystery Religion*, Ralph Woodrow Evangelistic Association, Riverside, California, 1966.

CHAPTER 7

Beastly Curiosities

This chapter contains no mysticism, only hard facts about 666 discovered by various mathematicians. They illustrate two things. The first is the quickness and fertility of mind of mathematicians. Mathematicians, whatever their faults, are *smart* people. The second is that the discoveries made by numerologists are not so astounding. Other people can make them too. The reason for this is the richness and depth of numbers: they contain an infinity of surprises, some delightful, some amazing, and some, as the following properties of 666, amusing. It is no wonder that some smart people should devote their lives to numbers. Numbers are worth it.

Magic squares have had a semimystical appeal ever since the discovery (when and by whom lost in the mists of history) of the smallest one, with magic sum of 15 along each row, down each column, and along the two main diagonals:

4	9	2
3	5	7
8	1	6

Albrecht Dürer, among other things a mathematician and hence a smart person, put this magic square

16	3	2	13
5	10	11	8
9	6	7	12
4	15	14	1

in the upper left-hand corner of his etching *Melancholia*, composed in 1514. Notice the two numbers in the middle of the last row of the square. Mathematicians, smart people, sometimes go in for understated virtuosity.

The mysticism of magic squares goes well with the mysticism of 666. Rudolf Ondrejka [7] found

$$\begin{array}{ccc} 232 & 313 & 121 \\ 111 & 222 & 333 \\ 323 & 131 & 212 \end{array}$$

a magic square with palindromic entries whose sum is the palindrome 666. He also constructed [6] a four-by-four magic square with sum 666:

$$\begin{array}{rrrr} 320 & 169 & 138 & 39 \\ 26 & 151 & 208 & 281 \\ 195 & 294 & 13 & 164 \\ 125 & 52 & 307 & 182 \end{array}$$

Not only do the rows, columns, and both main diagonals sum to 666, so do all the broken diagonals (e.g., 138 + 281 + 195 + 52), as do the nine 2-by-2 subsquares (169 + 138 + 151 + 208 = 666), the four corner squares of the 3-by-3 subsquares (169 + 39 + 294 + 164 = 666), the 4-by-2 rectangles $(26+281+195+164 = 666)$, and the 4-by-4 square $(320+39+125+182 = 666)$, or any square obtained from it by cyclically permuting columns, or rows. It is a *very* magic square.

Further, note that it contains the sinister 13, as well as $169 = 13 \cdot 13$, and has four arithmetic progressions with difference 13:

$$\begin{array}{l} 13, 26, 39, 52 \\ 125, 138, 151, 164 \\ 169, 182, 195, 208 \\ 281, 294, 307, 320. \end{array}$$

Mystics could, I think, contemplate it by the hour.

Not being sensitive to numerology, Dr. Ondrejka did not do what modern numerologists do, reduce the elements of the square modulo 9. Modern numerologists do not use "modulo 9" because they have never encountered the term. Instead, they tell you to add the digits of a number (169: $1+6+9 = 16$) and repeat the process (16: $1+6 = 7$) until a single digit remains. This process, once well known as "casting out nines," gives the remainder when an integer is divided by 9. Applied to the magic square, we get

$$\begin{array}{cccc} 5 & 7 & 3 & 3 \\ 8 & 7 & 1 & 2 \\ 6 & 6 & 4 & 2 \\ 8 & 7 & 1 & 2 \end{array}$$

More magic: there π is (3.141...), down the third column, and e (2.718...), a little scrambled, in rows two *and* four, and 666 replicated, almost, in row three (66[2 + 4]) as well as in column three ($3 \cdot 222$). Numerologists would suspect that Dr. Ondrejka put those features in his square, but I am sure he would be surprised to know about them. Numbers can do all sorts of fascinating things, all by themselves.

Here is an even bigger magic square, by Alan William Johnson, Jr. [3], with sum 666.

3	107	5	131	109	311
7	331	193	11	83	41
103	53	71	89	151	199
113	61	97	197	167	31
367	13	173	59	17	37
73	101	127	179	139	47

What is notable about this square is that it is pandiagonal, like the four-by-four square by Dr. Ondrejka, so all its broken diagonals in both directions sum to 666. Even more notable, its entries are distinct primes. This could be taken as having numerological significance, since the primes are the most deficient of numbers. (An integer is deficient if the sum of its divisors that are less than itself is smaller than the integer. That is, n such that $\sigma(n) - n < n$. For example, 15 is deficient because $1 + 3 + 5 < 15$. For primes, the sum is as small as it can possibly be: 1.) The beast, being evil, must be deficient, so it is fitting that the square sums to 666. Numerologists do not discover astonishing things like this. Reducing numbers modulo 9 seems to be about the limit of their abilities. Can it be that they are not as smart as mathematicians?

Everyone knows that a six-by-six magic square with entries $1, 2, \ldots, 36$ has the sum of all its entries equal to 666, since

$$1 + 2 + 3 + \cdots + 36 = (36 \cdot 37)/2 = 666.$$

Not everyone knows what Monte Zerger discovered [13], that if you make an eleven-by-eleven-by-eleven magic cube with entries $1, 2, \ldots, 1331$ (n-by-n-by-n magic cubes exist for all odd n from seven up), the number in the central cell must be 666.

This came from the problem of finding integer solutions of

$$n^3 + 1 = k^2(k^2 + 1).$$

The magic cube corresponds to the solution $n = 11$, $k = 6$. A computer search by C. R. J. Singleton has shown that there are no other solutions for

$k < 10{,}000{,}000$. It is likely that there are no suitable values of k other than 6, which would give mystical support to the uniqueness of the beast.

Everyone also knows that "Hitler" sums to 666 in the alphabet where a = 100, b = 101,..., z = 125. Not everyone knows what was discovered by A. A. Castro [2], that if we use the alphabet with a = 2, b = 3, c = 5,..., z = 101 (the first 26 primes), then "Hitler" is $19+23+71+37+11+61 = 222$. "Another coincidence?" asks the author. I know of no one else who has thought of using the primes in this way. But, primes being maximally deficient, it is an appropriate alphabet for beasting. It is slightly embarrassing that 666 is not also deficient, but there is no getting around

$$\sigma(666) = \sigma(2 \cdot 3^2 \cdot 37) = \sigma(2)\sigma(3^2)\sigma(37) = 3 \cdot 13 \cdot 38 = 1482 > 1332$$

The excess, $1482 - 1332 = 150$, does not seem to be numerologically significant.

Another connection between 666 and primes (I do not know who first observed it) is that 666 is the sum of the squares of the first $6 + 6/6$ primes:

$$666 = 2^2 + 3^2 + 5^2 + 7^2 + 11^2 + 13^2 + 17^2.$$

The six theme is not hard to exploit [12]:

$$\begin{aligned} 666 &= 6 + 6 + 6 + 6(6 \cdot 6 + 6 \cdot 6 + 6 \cdot 6) \\ &= 6 + 6 + 6 + 6 \cdot 6(6 + 6 + 6) \\ &= 6^4 - 6^3 - 6^3 - 6^3 + 6 + 6 + 6 \\ &= 6 \cdot 6(6 \cdot 6 - 6 - 6 - 6) + 6 + 6 + 6. \end{aligned}$$

The beast has some connections with the Fibonacci sequence [10]. If f_n denotes the nth Fibonacci number,

$$f_{n+1} = f_n + f_{n-1}, \qquad \text{with } f_0 = 0, f_1 = 1,$$

then

$666 = f_1^3 + f_2^3 + f_4^3 + f_5^3 + f_6^3$: the sum of the subscripts is $6 + 6 + 6$.

$666 = (f_1^3 + (f_2 + f_3 + f_4 + f_5)^3)/2 = f_6^2 + f_7^2 + f_8^2 - f_3^3$; in the last expression, the sum of the subscripts (with proper sign—$6 + 7 + 8 - 3$) is again $6 + 6 + 6$.

$$666 = f_{15} + f_{11} - f_9 + f_1, \quad \text{and} \quad 15 + 11 - 9 + 1 = 6 + 6 + 6.$$

Non-Fibonacci but impressive identities include

$$666 = 2^9 + 2^7 + 2^3 + 6 + 6 = 3^6 - 3^4 + 3^3 - 3^2,$$

$$666 = 18 \cdot 37; 37 = 19 + 18;$$

$$666 = 37^2 - 19^2 - 18^2 - (6 + 6 + 6),$$

and

$$666 + 6 + 6 + 6 = 5^3 + 6^3 + 7^3.$$

To make that last identity more impressive, I could write 5 as $6 - 1$, 7 as $6 + 1$ (and, for that matter, 1 as $6/6$) and 3 as $(6 + 6 + 6)/6$. In fact, I think that I will:

$$666 + 6 + 6 + 6 = \left(6 - \frac{6}{6}\right)^{(6+6+6)/6} + 6^{(6+6+6)/6} + \left(6 + \frac{6}{6}\right)^{(6+6+6)/6}$$

Notice that the number of 6s on the right-hand side is $6 + 6 + 6 + \frac{6}{6}$.

Michael Keith [4] noted a number of properties of 666.

666 is the 36th (6×6) triangular number. It is the largest triangular number whose decimal representation has a single digit [1].

$666 = 18 \cdot 37$, $1/37 = .021021021\ldots$, and $-(1 \cdot 8) + (3 \cdot 7) = 13$. What of it, you ask? You see the $21 = (3 \cdot 7)$, 13, and 18, so intertwined in 666? The Bible verse mentioning the beast is in the twenty-first book of the New Testament, chapter 13, verse 18.

The next verse (Revelation 14:1) mentions the number 144,000. Notice that $144000/666 = 216.216216216\ldots$ and $216 = 6 \cdot 6 \cdot 6$.

$$\varphi(666) = 216 = 6 \cdot 6 \cdot 6.$$

($\varphi(n)$ is Euler's φ-function of n: the number of positive integers less than n and relatively prime to n. $\varphi(15) = 8$ because the integers from 1 to 15 that have no factor in common with 15 are 1, 2, 4, 7, 8, 11, 13, and 14.)

$$1^6 - 2^6 + 3^6 = 666.$$

$$666 = 6 + 6 + 6 + 6^3 + 6^3 + 6^3.$$

There are only five other integers like that, equal to the sum of their digits and the cubes of their digits: 12, 30, 870, 960, and 1998. The proof of that is not hard. For five-digit numbers, the sum of the digits and their cubes falls short—the biggest it can get is $5 \cdot 9 + 5 \cdot 93 = 3690$, only four digits long. For six-digit or larger integers, it falls even shorter. So, only 10,000 cases remain, and checking them is the sort of thing that computers love to do. Or that computer users love to do: the problem is easily programmed and the solution is satisfying when complete. The program would find the last example when it calculated that

$$1 + 9 + 9 + 8 + 1 + 729 + 729 + 512 = 1998.$$

The numerological implication of 1998 is clear: $1998 = 666 + 666 + 666$. We should, I think, be very careful in 1998.

Note the mathematical and numerological possibilities here opened up. What if, instead of cubes, we take squares, or fourth powers? What if, instead of base 10, we write numbers in base 6, the beastly base, or the doubly beastly base 12? Since numerologists are not going to look into them, someone else will have to do the job for them.

If we search in π, we find the 666th, 667th, and 668th digits after the decimal point are $343 = 7 \cdot 7 \cdot 7$.

C. Singh [11] has shown that Ramanujan's number,

$$1729 = 10^3 + 9^3 = 12^3 + 1^3,$$

the smallest integer that is a sum of two cubes in two different ways, is, for some reason, connected with 666:

$$1729 = (6 + 1)(6 + 6 + 1)(6 + 6 + 6 + 1).$$

$$\text{Permute 1729:} \quad 9721 - 1729 = (6 + 6)(666).$$

$1729 + 666 = 2 + 3^2 + 5^2 + 7^2 + 11^2 + 13^2 + 17^2 + 19^2 + 23^2 + 29^2$, the squares of all the primes up to 29, though 2, the oddest prime (because it is even) appears only to the odd power 1.

$$666 = 7 \cdot 7 + 13 \cdot 31 + 17 \cdot 71 - 19 \cdot 91 + 23 \cdot 32.$$

That is striking, as is $19 \cdot 91 = 1729$.

Clifford Pickover [9] noticed that the 3184th Fibonacci number, and the four after it, have 666 digits. Dr. Pickover did not notice that

$$3^3 + 1^3 + 8^3 + 5^3 = 665$$

and adding 1 (since 3185 is one more than 3184, the position of the first beastly Fibonacci number) gives... guess what?

A curiosity is ([5])

$$\underset{6}{1 + 2 + 3} = \underset{6}{1 \cdot 2 \cdot 3} = \underset{6}{\sqrt{1^3 + 2^3 + 3^3}}$$

An amazing curiosity is ([8])

$$5(55)(555) = 152625.$$

That alone is not amazing, but if you increase each digit by 1, the equation is still true:

$$6(66)(666) = 263736.$$

There are no other numbers like that.

The conclusions to draw from these curiosities are that mathematicians are clever, that numbers are wonderful, and that 666s are easy to find if you know how to look. So, when a beaster comes at you waving his 666, you can yawn, say "That's nothing—look at *this*," and show him one or more of the properties of 666 in this chapter. It won't do any good, but it will give you something to do.

References

1. Ballew, D. W. and R. C. Weger, Repdigit triangular numbers, *Journal of Recreational Mathematics* **8** (1975–76) #2, 96–98.
2. Castro, Almerio Amorim, Letter to the editor, *Journal of Recreational Mathematics* **16** (1983–84) #4, 249.
3. Johnson, Allan William, Jr., Letter to the editor, *Journal of Recreational Mathematics* **16** (1983–84) #4, 247.
4. Keith, Michael, The number 666, *Journal of Recreational Mathematics* **15** (1982–83) #2, 85–87, 122.
5. Moessner, Alfred, Curiosum, *Scripta Mathematica* **13** (1947), 57.
6. Ondrejka, Rudolf, Letter to the editor, *Journal of Recreational Mathematics* **16** (1983–84) #2, 121.
7. Ondrejka, Rudolf, Problem 1641. *Journal of Recreational Mathematics* **20** (1988) #2, 150. Solution, **21** (1989) #2, 154–156.
8. Penning, P., Problem 1620, solution by Sam Baethage, *Crux Mathematicorum* **18** (1992) #2, 61–62.
9. Pickover, Clifford A., Apocalypse numbers, *Mathematical Spectrum* **26** (1993–94) #1, 10–11.
10. Singh, Chanchal, The beast 666, *Journal of Recreational Mathematics* **21** (1989) #4, 244.
11. Singh, Chanchal, More on 1729, *Journal of Recreational Mathematics* **21** (1989) #2, 135–136.
12. Trigg, Charles W., The perfectly beastly number, *Journal of Recreational Mathematics* **20** (1988) #1, 61.
13. Zerger, Monte, Problem 1989. *Journal of Recreational Mathematics* **25** (1993) #3, 237.

CHAPTER 8

The Beast is Coming!

When Your Money Fails, by Mary Stewart Relfe, Ph.D. [1], subtitled "the '666 system' is here" is an example of what 666 can do to people, and what people can do to 666. Dr. R., who nowhere discloses the field of her doctoral degree, has noticed the number in many places:

> It was not the design of the floor tile made here in the United States and purchased locally in Montgomery, Alabama, that almost mesmerized me, it was the bold prefix "666" stamped on both sides! I had just filed a photograph of a man's dress shirt manufactured in China and purchased in the United States with the number "666" on the label. Before closing the folder, I gazed at the 8 × 10 glossy which AP Wire Photo Service had just mailed to me of the official reopening of the Suez Canal showing the first warship entering the canal carrying on board Egyptian President Anwar Sadat, which had on its bow the big bold numbers "666." My repertoire of recent information concerning the national and international usage of the number "666" was becoming engorged.
>
> The file contains additional information which indicates that:
>
> - World Bank code number is "666."
> - Australia's national bank cards have on them "666."
> - New credit cards in the U. S. are now being assigned the prefix "666."
> - Olivetti Computer Systems P6060 use processing numbers beginning with "666."
> - Central computers for Sears, Belk, J. C. Penney and Montgomery Ward prefix transactions with "666" as necessitated by computer programs.
> - Shoes made in European Common Market Countries have stamped on inside label "666."
> - Visa is 6 6 6; Vi, *Roman* numeral, is 6; the "zz" sound, Zeta, the 6th character in the *Greek* alphabet, is 6; a, *English* is 6.

The equivalence of "a" to 6 in the above beasting of "Visa" may not be clear. Later, Dr. R. uses the alphabet with a = 6, b = 12, c = 18, . . . , z = 156 to point out that using it "computer" gematrizes to 666, as does "Kissinger." I think that is where she gets that value for "a." A different alphabet would be needed to beast "MasterCard."

- Computers made by Lear Sigler have a seal on the side on which is stamped the number "666."
- IRS Alcohol, Tobacco, and Firearms Division has on their employee badges the number "666."
- IRS Instructions for Non-profit Corporation Employee 1979, W-2 Form requires the prefix "666."
- IRS began to require the prefix "666" on some forms; for example, W-2P, disability is 666.3; death is 666.4, etc., as early as 1977.
- State Governments are now using on their office purchasing paperwork the number "666."
- President Carter's new Secret Security Force patches have on them "666."
- The McGregor Clothing Company recently introduced its new "666" Collection of menswear.
- A midwestern telephone company's credit card is encoded "666."
- Identification tags on all foreign made Japanese parts for the Caterpillar Company, Peoria, Illinois, contain the code "666."
- Work gloves manufactured by the Boss Glove Company are stamped "666."
- The Crow's Hybrid Corn Company of Nevada, Iowa, offers a "666" seed as its top yielding hybrid. [1, pp. 15–18]

I could go on. There are eighteen (6 + 6 + 6) more instances of 666, including

- Elementary algebra book is entitled "666 Jellybeans."

However, you probably have the idea by now. 666s are everywhere, and they are spreading.

> When my friend and colleague Sally O'Brien received her J. C. Penney statement for August, 1980, no one had consulted with her or notified her that the number on her old card was obsolete, nor have they yet issued a new card. Her statement, however, reflected her account number was changed from 516-747-847-7-2, to 666-742-522-42. Further examination of her 1978 Sears credit card, revealed that it incorporated this code into a suffix "666." See illustration. I facetiously said, "Sally, you are already a part of the Antichrist's system." She quickly responded, "As soon as I get back to California, I am going to draw out of my savings, pay this account off, and get out of the system." [1, p. 60]

As the mention of the antichrist shows, all the 666s are signs of the end times, precursors of the long-awaited winding up of the world. More specifically, Dr. R. asserts that they are all part of a satanic plot to replace cash with cards, specifically a universal card to be used for identification and for all monetary transactions.

Actually, the 666s are manifestations of the workings of the Law of Small Numbers, considered in more detail in chapter 9. The Law states that there are not enough small integers for the many duties that they must perform. Thus there will be many 666s in circulation, just as there are many 667s and 665s. If the Boss Glove company gave its line of gloves some number other than 666, then another company, making something else, could be found putting 666 on its products. For a sufficiently large monetary consideration, I will produce a list as sinister and impressive as Dr. R.'s for *any* three-digit number. You name the number. The Law of Small Numbers guarantees that I can make the list.

Dr. R.'s publisher was evidently a little nervous about her implication that many large and powerful organizations were in league with the devil, because on the title page the following statements were included:

> The Publisher endeavors to print only information from sources believed reliable, but absolute accuracy cannot be guaranteed.

and

> Information in this book is not an indictment against any product, person, or institution, financial or otherwise, but simply evidence of Bible Prophesy being fulfilled in this final World System.

When the universal card comes to pass, Dr. R. says

> I expect the code sequencing to be thusly:
>
> 666 the International Code which will activate the World Computer.
>
> 110 the National Code which will activate the Central United States computer.
>
> 203 or, your area telephone code, indicating your locality.
>
> Your nine digit Social Security Number.
>
> This construction will make my own personal number to be:
>
> $$\frac{666\ 110}{6} + \frac{205\ 419}{6} + \frac{386\ 968}{6}$$
>
> Three six-digit units! [1, p. 55]

Diehards who refuse to give up their cash will be taken care of by worldwide inflation, every bit as severe as Germany's after the first world war, making all currency valueless.

> My Bible has 1500 pages in it. Can you imagine having to count each page of a Bible to purchase a loaf of bread? The Germans had to count 268,000 marks

> (if they were all in one million denominations)! You can see how Christians who do not submit to the Electronic Money System (see that chapter) will be ostracized in times of hyperinflation. Think of the convenience of a single card and the inconvenience of hauling millions of dollars to the market place for a loaf of bread, where the merchant will resent having to assist you.
> [1, pp. 98–99]

After, or with, the card will come marks. Not German marks, but physical ones, like tattoos. This so as to agree with the two verses in Revelation (13: 16–17) that come just before the mention of 666. In the King James version:

> And he causeth all, both small and great, rich and poor, free and bond, to receive a mark in their right hand, or in their foreheads.
>
> And that no man might buy or sell, save he had the mark, or the name of the beast, or the number of his name.

The mark, Dr. R. surmised, will be the Universal Product Code (UPC) because of the 666 built into it (see chapter 7) and she includes a scary drawing [1, p. 32] of a person whose forehead is almost filled with a UPC, with a 666 among its numbers. She found an example of a UPC that had a horizontal F and a horizontal H on its bottom.

> Conjecture: when mark is inserted on body, F = Forehead, H = Hand location.
> [1, p. 30]

The advantage of the mark is that it cannot be lost or, presumably, counterfeited, and thus will make easier the task of the world dictator to impose his will on everyone. As evidence of the coming of the marks the author has two citations. The first is from "*The Cronkhites*, Polson, Montana."

> "The Internal Revenue Service refund office has printed a trial run of their checks for 1984 and twenty-five accidentally got mailed recently in California." On the back of the checks are these instructions: "Do not cash this check unless the recipient has a number on either his right hand or on his forehead."
> [1, p. 58]

The second is from "The Scroll, August, 1980, by Evangelist Darrell Dunn."

> In addition, in July and August, 1980, the Internal Revenue Service mailed scores of Social Security checks to recipients with instructions on the back being: "*The proper identification Mark on the right hand or forehead.*" Upon the banks' refusal to cash these checks amidst much confusion and denials, the IRS admitted their mistake thusly "THESE GOVERNMENT CHECKS REQUIRING A MARK IN A PERSON'S RIGHT HAND OR FOREHEAD ARE NOT TO BE PUT INTO USE UNTIL 1984."
> [1, p. 58]

Leaving aside the detail that the IRS does not distribute social security checks, these vaporings are, literally, incredible. They verge on insanity. They *are* insane.

But Ms. R., Ph.D., evidently credits them, and she is not insane. There is a picture of her on the inside front cover of her book, and she appears as sane as you or I. She looks as if she could be a successful seller of real estate. Her grin seems a bit forced, and there does seem to be something a little funny about her eyes, but I may think that only because I know she wrote her book. A crazed person is not likely to be

> A successful businesswoman, her interests include a multi-million dollar health care facility in downtown Montgomery, farming, leasing, and furniture. She was Montgomery's Woman of Achievement for 1975, and nominated for Who's Who in 1977.
>
> She is a Commercial Pilot, Multi-Engine Instrument Flight Instructor, and has piloted her plane on international flights. A two-term Commissioner in the Alabama Aeronautics Department, she is now in her third term as Secretary-Treasurer of the Montgomery Airport Authority.

Away from 666, she does well, obviously functioning with a clear head. But when numbers get into heads, funny things can happen:

> More recent information has surfaced that the banking industry in conjunction with the United States Government has completed testing of a soft injectable plastic manufactured by a laboratory in Orlando, Florida. Upon a subcutaneous injection of the liquid substance, it smooths out under the skin much like water does on a flat surface; becomes semi-hard, forming a permanent under-skin shield on which the tattoo gun imprints the person's number. Further information indicates that banks are now ready to begin imprinting numbers "*in* the right hand or *in* the forehead." Revelation 13:16. Note that John said "IN" not "ON!" [1, p. 44]

After we all have marks implanted in us the world dictator will take over and the gaudy events leading to the end of the world will get into full swing. Her hints that 1987 would be the year were guarded, as turned out to be wise. Her identification of Anwar Sadat with the antichrist was also tentative:

> **I maintain Sadat is History's nearest prototype of the Jewish False Messiah!** [1, p. 138]

Her book has evidently been a success, since its publisher labels it "#1 International Bestseller" and claims that 600,000 copies were sold in five months. Even discounting that by a factor of ten, there are some 66,666 people who have read about how our government is going to give us all unremovable marks on, or in, our foreheads, and some of them no doubt believed it. It's in a book, after all, and one by a Ph.D. The book has spawned a sequel, *The New Money System* that, according to an advertisement, is all about the marks that will replace money. It too contains 666s.

Dr. R. is not an example of a head turned by the power of numbers. She is an example of a head already turned, appealing to the power of numbers to turn other's heads and make them see the world as she does. The point of converting people to premillennarianism is not entirely clear to me—I suppose it is that they can repent of their sins before it is too late. But what a disappointment the end-of-the-worlders will all have as the year 2000 slips into history with the world still more or less intact! It will be *such* a wait until 3000.

Reference

1. Relfe, Mary Stewart, Ph.D., *When Your Money Fails*, by Mary Stewart Relfe, Ph.D., Ministries, Inc., Montgomery, Alabama, 1981.

CHAPTER 9

The Law of Small Numbers

Richard K. Guy, currently of the University of Calgary, is a mathematician of power, erudition, and wit. In one of his many publications (one of his lesser works) he points out that there are exactly 10! seconds in six weeks [3]. Another contribution to mathematics, and one of more importance, is his enunciation of the Strong Law of Small Numbers [1]:

> There aren't enough small numbers to meet the many demands made of them.

How true! As we will see, the Law has applications to numerology as does one of its corollaries, the Law of Round Numbers, considered in chapter 11.

The name of the Law comes from the Law of Large Numbers of probability which states, roughly, that in the long run, random events behave as expected. Toss a coin many times and the proportion of heads will get close to one-half. Roll two dice a few trillion times and the number of sevens will be close to one-sixth of the number of tosses. The Weak Law of Large Numbers says that, for the coin example,

$$\lim_{n\to\infty} P\left\{\left|\frac{\text{number of heads in } n \text{ tosses}}{n} - \frac{1}{2}\right| < \epsilon\right\} = 1$$

no matter how small ϵ is. The Strong Law says that the difference between the proportion of heads and $1/2$ will be less than ϵ for all sufficiently large n almost certainly.

The Law of Small Numbers has nothing to do with probability, except that it implies that, since there are too few small integers to go around, coincidences will occur that can mislead. About the Law, Professor Guy wrote in 1994:

> I claim it as my own, though really only the name, because it's a phenomenon that most mathematicians have observed. Andrew Granville recently sent me a quote from Fermat, writing about Wallis, which really contains the germ

> of the idea. It slowly developed into articles after discussions with Erdős and many others.
>
> It was really a question of finding a name for the frustration that one often felt in combinatorics and number theory in not knowing whether to go for a proof of a theorem or to continue searching for a counterexample. I think it began to crystallize about thirty-five years ago. I asked Lowell Schoenfeld if he believed the Riemann Hypothesis. He said "yes" so I asked him why he was looking for zeros. He said that he had no hope of proving the hypothesis but there was always a chance, however slim, that he could find a counterexample.

Professor Guy says that the Law is an enemy of mathematical discovery, since

> superficial similarities spawn spurious statements

and

> capricious coincidences cause careless conjectures.

It can also work the other way, because

> early exceptions eclipse eventual essentials

and

> initial irregularities inhibit incisive intuition.

Apt alliteration's artful aid notwithstanding, an example is given by alternating sums of factorials:

$$3! - 2! + 1! = 5$$

$$4! - 3! + 2! - 1! = 19$$

$$5! - 4! + 3! - 2! + 1! = 101$$

$$6! - 5! + 4! - 3! + 2! - 1! = 619$$

$$7! - 6! + 5! - 4! + 3! - 2! + 1! = 4421$$

$$8! - 7! + 6! - 5! + 4! - 3! + 2! - 1! = 35899$$

The sums, 5, 19, 101, 619, 4421, and 35899 are all primes. The question is, can this go on? Anyone with experience in the ways of prime numbers would conclude that it could not, and that would be correct since the next case is

$$9! - 8! + 7! - 6! + 5! - 4! + 3! - 2! + 1! = 326981 = 79 \times 4139.$$

Among small integers, primes are common so it is not a surprise that a collection of eight odd integers happens to contain only primes. It is not likely, but it is not a surprise. The chance that eight odd integers selected at random from small integers like those in the example are all prime is something like $(.3)^8 = .00006561$. This is quite small, but when you consider the vastness of the universe of eight-integer sequences, it is inevitable that occasionally a sequence with nothing but primes in it will occur. Coincidences *must* happen. This is a fact that numerologists often choose to ignore.

Here is a more surprising example. Multiplication, tedious to do by hand, quicker with a computer algebra system, shows that

$$(x+y)^3 = x^3 + y^3 + 3xy(x+y)(x^2+xy+y^2)^0$$

$$(x+y)^5 = x^5 + y^5 + 5xy(x+y)(x^2+xy+y^2)^1$$

$$(x+y)^7 = x^7 + y^7 + 7xy(x+y)(x^2+xy+y^2)^2$$

It certainly looks as if something is going on there, and even those who are on intimate terms with polynomials might give in to the temptation to conjecture:

THEOREM: $(x+y)^{2n+1} =$

$$x^{2n+1} + y^{2n+1} + (2n+1)xy(x+y)(x^2+xy+y^2)^{n-1},$$

$n = 0, 1, 2, 3, \ldots$

The temptation to start writing a proof would be hard to resist as well:

Proof: We will use the method of mathematical induction. The theorem is true for $n = 1, 2, 3$. Suppose that it is true for $n = k$. Then we have

$$\begin{aligned}(x&+y)^{2k+3}\\ &= (x+y)^2(x+y)^{2k+1}\\ &= (x+y)^2(x^{2k+1} + y^{2k+1} + (2k+1)xy(x+y)(x^2+xy+y^2)^{k-1}).\end{aligned}$$

But multiplying the terms out and doing algebra would not result in success because the theorem is false for $n = 4$. The quick way to see this is to put $x = y = 1$ in the theorem to see that it asserts, in particular, that

$$2^{2n+1} = 1 + 1 + (2n+1) \times 2 \times 3^{n-1}.$$

That is, $2^{2n} - 1 = (2n+1)3^{n-1}$. It is just luck that this is true for $n = 1, 2$, and 3. The left-hand side is approximately 4^n while the right-hand side is a bit bigger than $2n \cdot 3^n$, so eventually the left-hand side is going to be larger than the right-hand side. In fact, it happens at $n = 4$, since $255 > 9 \cdot 27 = 243$.

In that example, the operation of the Law of Small Numbers did not lead to the waste of mathematical time, since the first counterexample came so soon. In the next example, it could, and perhaps has, led to time spent trying to prove the unprovable. Consider the sequence defined by $x_0 = 1$ and

$$x_{n+1} = \frac{1 + x_0^2 + x_1^2 + \cdots + x_n^2}{n+1}$$

It starts

n	0	1	2	3	4	5	6	7
x_n	1	2	3	5	10	28	154	3520

The question is, is x_n always an integer? The terms of the sequence quickly get large and the next two at least do not give a counterexample ($x_8 = 1551880$, $x_9 = 267593772160$). There are other sequences, similarly defined by equations that have numerators and denominators, that have only integer values. The urge to try to prove that x_n is an integer for all n could very well be too strong to resist, and a good deal of time and paper might be spent in the effort. But no matter how much time and paper were invested, no proof would result. The Law of Small Numbers is at work. Although x_n is an integer for $n = 0, 1, \ldots, 42$, x_{43} is not.

These examples are all taken from [1]. There are many more there, as well as in its sequel [2].

An example of a misleading initial irregularity arises in the problem of representing integers as sums of powers. As was proved in the eighteenth century, every integer is a sum of four (or fewer) squares, for example,

$$19 = 4^2 + 1^2 + 1^2 + 1^2$$

$$20 = 4^2 + 2^2$$

$$21 = 4^2 + 2^2 + 1^2$$

and so on. Further, there are infinitely many integers that need four squares. Any integer of the form $8n + 7$ cannot be written as a sum of only three squares—31 is a sum of four squares

$$31 = 5^2 + 2^2 + 1^2 + 1^2 = 3^2 + 3^2 + 3^2 + 2^2,$$

but no fewer than four will do.

When we come to cubes, looking at small integers we might guess (as Edward Waring did in 1770) that every integer is a sum of nine or fewer cubes. That nine is necessary is shown by

$$23 = 2^3 + 2^3 + 1^3 + 1^3 + 1^3 + 1^3 + 1^3 + 1^3 + 1^3.$$

This is a mere early exception. 239 is another, but it is the *last* integer to need as many as nine cubes. It is very likely that 454 is the last integer that needs eight cubes ($1 \cdot 7^3$, $4 \cdot 3^3$, and $3 \cdot 1^3$) and 8042 is the last that needs seven:

$$8042 = 16^3 + 12^3 + 10^3 + 10^3 + 6^3 + 1^3 + 1^3.$$

Once we get past the initial irregularities, it looks fairly certain (though it has not been proved yet) that every integer can be expressed as a sum of six or fewer cubes.

The problem for fourth powers is easier than for cubes. 79 takes nineteen fourth powers, but once we get beyond the early exceptions, sixteen fourth powers are enough to represent any number.

This is not how the Law can be applied to numerology, because numerologists seldom deal with integers with more than five or six digits, and most of them restrict themselves to two or fewer. Even 999999 is a small integer as are, for that matter, 999999999 and 999999999999. Since the human race has written down only finitely many integers in its finite lifetime, no one has ever seen a *really* large integer. (It is possible to claim that

$$((((10^{100})!)^{100})!)^{100}$$

is a large integer but, first, there are infinitely many integers which are larger and only finitely many which are smaller and, second, an integer whose tenth digit you cannot determine is not one that you have actually *seen*.)

The application of the Law to numerology is that, since there are so few small integers, coincidences will occur. Non-numerological coincidences occur all the time and we think nothing of them. For example, here are a few threes, many from [5]:

Christian creeds (Apostles', Nicene, and Athanasian)
ships of Christopher Columbus (Niña, Pinta, and Santa Maria)
Fates (Clotho, Lachesis, and Atropos)
fiddlers (in "Old King Cole")
Furies (Alecto, Megaera, and Tisiphone)
Bears (Mama, Papa, and Baby)
Graces (Aglaia, Euphrosyne, and Thalia)
goals in a hat trick (in ice hockey)
Rs (reading, writing, and arithmetic)
subjects in the trivium (grammar, rhetoric, and logic)
Kingdoms (animal, vegetable, and mineral)
Magi (Gaspar, Melchior, and Balthazar)
Marx Brothers (Chico, Harpo, and Groucho)
degrees of burns (first, second, and third)
estates (nobility, clergy, commonality)
Reichs (First, Second, and Third)
bags of wool (one for my master, one for my dame, one for the little boy who lives down the lane)

dimensions (length, width, and height)
Musketeers (Athos, Porthos, and Aramis)
Stooges (Larry, Moe, and one other)

There are other threes that could be mentioned, such as faith, hope, and charity; red, white, and blue; body, mind, and spirit; and Shadrach, Meshach, and Abednego. What do we make of all these threes? Nothing, of course. They have no deep significance because they have no significance at all. We smile at the variety of things that come in threes and, when we do, we are unconsciously applying the Law of Small Numbers. There are so many things in the world that it is no surprise that many of them come in threes.

Number mystics, though, think that there are *reasons* why things come in threes. Mystical reasons, to be sure, but reasons nevertheless. They cannot accept that, because of the Law of Small Numbers, some sets have three elements because they have to have *some* small integer of members. The Three Musketeers could very well have been four, by counting d'Artagnan. There once were five Marx Brothers (Gummo and Zeppo dropped out). Chance, that's why there are three of each.

Even nonnumerologists can teeter on the edge of seeing patterns. Annemarie Schimmel had a success with *The Mystery of Numbers* [4], a title echoing *De numerorum mysteriis* of Peter Bungus (1583), a pioneering two-volume work on number symbolism. She is no numerologist, but a specialist in Islam. In her chapter on three, we have

> For Wolfgang Philipp, all being consists of a tripolar *Ergriffenheit* (emotion), which is manifested in wave, radiation, and condensation, and he thinks that because we are existentially tripolar, we feel at home in corresponding triads. That is why three things are good things, and we are in a good mood when we find our own active, middle and passive principles fulfilled and confirmed in them.
>
> In 1903 the German scholar R. Müller tried to explain the importance of 3 in tales, poetry, and visual art and argued that the importance of the triad stems from the observation of nature. Once human beings saw water, air, and earth, they developed the idea of the existence of 3 worlds (called Midgard, Asgard, and Niflheim in the Germanic tradition); they recognized 3 states (i.e. solid, liquid, and gaseous); they found 3 groups of created things (minerals, plants, and animals) and discovered in the plants root, shaft, and flower, and in the fruit, husk, flesh, and kernel. The sun was perceived in a different direction and form in the morning, at high noon, and in the evening. In fact, since the world we see and live in is 3-dimensional, all our experiences take place within the coordinates of space (length, height, and width) and time (past, present, and future). All of life appears under the threefold aspect of beginning, middle, and end, which can be expressed in more abstract terms as becoming, being, and disappearing; a perfect whole

> formed of thesis, antithesis, and synthesis. There are also the 3 primary colors, red, yellow, and blue, from which all other colors can be mixed.
> [4, pp. 59–60]

Does it not seem as if three is in control? That is, that it is the number that is coming first and drawing the examples to itself, rather than having non-numerical reality force triads to our attention?

> In philosophy and psychology, 3 serves as the number of classification: time, space, and causality belong together. Since Plato, the ideal has been taken to be composed of the good, the true, and the beautiful, while Augustine established the categories of being, recognizing, and willing. The Indian *Chandogya Upanishad* likewise mentions several triadic groups, such as hearing, understanding, and knowledge, and in the later Upanishads, the 3 basic values that express the fullness of the one divine being are *sat*, *chit*, and *ananda* (being, thinking, and bliss). [4, p. 65]

There exist many examples of items that can be classified in three ways. That is a matter of fact, about which there can be no argument. To go from that to the statement that three *is* the number of classification is to leap into number mysticism. There is no arguing about that either, since mystical insights are personal. But the minute someone classifies something new in three ways because three is the number of classification and things must be classified in three ways, it becomes numerology.

In any event, that things can be classified in threes should not lead to any mystical insights. It is a only an instance of the Law of Small Numbers. Humans like to classify, and of the enormous number of things they have classified, many will be divided into three parts. A search would no doubt disclose almost as many things divided into four parts: the four seasons, the four temperaments, the four winds, the four elements (earth, air, fire and water—notice that fire was omitted in the quotation about Müller's opinions, an example of how the number three became more important than the reality), the four liberal arts of the quadrivium, and so on. There are also instances of five, six, and even nine being numbers of classification. There have to be. If you put ten million balls at random into one hundred boxes, every box is going to contain quite a few balls, which is another way of expressing the idea behind the Law of Small Numbers.

Remember the Law. It will be referred to again in this book, and you may be able to observe it operating elsewhere.

References

1. Guy, Richard, The strong law of small numbers, *American Mathematical Monthly*, 95 (1988) #8, 697–712.

2. ———, The second strong law of small numbers, *Mathematics Magazine*, 63 (1990) #1, 3–28.
3. ———, Did you know, *Crux Mathematicorum* 19 (1993) #3, 278.
4. Schimmel, Annemarie, *The Mystery of Numbers*, Oxford University Press, Oxford, 1993, originally published as *Das Mysterium des Zahl*, Eugen Diederichs Verlag, Munich, 1984.
5. Urdang, Lawrence, *The Facts on File Dictionary of Numerical Allusions*, Facts on File, 1986, reprinted as *Three-Toed Sloths and Seven-League Boots*, Barnes and Noble, New York, 1982.

CHAPTER 10
Comes the Revolution

The argument put forth again and again by pyramidologists, Stonehengers, Bible-searchers, and others who examine their chosen objects and find numbers in them, is that what they have found cannot have occurred by chance alone. There are just too many 7s and multiples of 7, or squares, or approximations to π, or whatever it is that they have found. A higher intelligence is involved, or a deity, or a secret group—*someone* had to put them there.

They fail to give chance sufficient credit. Amazing things can happen at random, but people are very reluctant to accept that. They insist that events must have *causes*, even when they don't. When children trip on rocks and hurt themselves, do they not sometimes give the rock a whop, saying "Bad rock!"? Wills, children think, are everywhere, even in inanimate objects, and wills exert themselves. Nothing happens by chance.

That is not the case. Consider the well-known experiment, never as far as I know actually carried out, of giving 1200 or so people a coin, telling them to flip it, keeping those people who flip a head and sending the rest home, and then repeating the process. After one flip, there will be roughly 600 survivors, after two around 300, and so on: the number of head-flippers will be cut approximately in half each time. After ten repetitions, there should be one flipper remaining who had tossed ten heads in a row, a remarkable exhibition of head-tossing ability. Would he not himself think that he had a rare talent? Would he not give interviews to the media, explaining his training methods and providing quotes like "Yes, Dan, before the last toss I was tense, very tense, but I could feel the head-energy flowing through me and while the coin was in the air I was completely calm"?

Of course, head-energy had nothing to do with it. The operation of chance—blind, purposeless chance, bestowing punishments and rewards at random, with never a thought or a care for humanity—guaranteed that the lucky person would exist.

By the way, real interviews like that imaginary one occur constantly. Mutual-fund managers, explaining how their funds have beaten the market for ten years in a row, never say "Well, I guess I was just lucky." They have methods. They have skills. They have reasons. Next year they will also have reasons to explain why they did not repeat their success. Baseball players and other athletes constantly give similar interviews, as do people who live to the age of 100. Luck is never a factor, even when it in fact is a major factor.

Sometimes the number-hunters will give elaborate computations to demonstrate the improbability that what they have found could have occurred by chance. For example, see the labors of Ivan Panin in chapter 12. "What they have found" is the key phrase: it is no fair applying probability to events that have already happened. I have a dollar bill whose serial numbers (and letters) are F 84323030 C. The chance that I should have that bill is

$$\frac{1}{12}\cdot\frac{1}{10}\cdot\frac{1}{10}\cdot\frac{1}{10}\cdot\frac{1}{10}\cdot\frac{1}{10}\cdot\frac{1}{10}\cdot\frac{1}{10}\cdot\frac{1}{10}\cdot\frac{1}{25}$$

($1/12$ because the first serial letter tells which of the twelve Federal Reserve Districts a bill comes from, and $1/25$ instead of $1/26$ for the final letter because I think that *O* is not used as a serial letter.) That is one in thirty billion, which is very small, too small, a number-hunter would say, to have happened by chance. It follows that some outside force must have seen to it that I should have that particular dollar bill.

It does not follow. There is no "therefore" there. The probability that I possess that dollar bill is not $.00000000003333\ldots$, it is 1, because I in fact have it. No outside forces are involved, besides the force of chance. Similarly, the chance that someone exists with exactly the genetic code that you have in your chromosomes is as close to zero as makes no difference—much smaller than one in a mere thirty billion—but I do not conclude that you are so improbable that mysterious outside forces must be responsible for your existence. Your existence is no more mysterious than mine, or than that of any of our billions of fellow-humans. We are here, and for no other reason than chance.

Given something that exists and has any complexity to it, the ingenious minds of humans can find all sorts of amazing things in it, whether they were put there on purpose or not. An example is given by a miniature pamphlet measuring 2.5″ by 3.5″, *History Computed*, by Arthur Finnessey [1]. Mr. F., browsing through an encyclopedia, was struck by the fact that

> Four of the first six U.S. Presidents were inaugurated at age 57. On a presidential data chart in an encyclopedia the four 57's stand out at the top of their column and surprisingly there is not another age 57 in the remainder of the long list. [1, p. 1]

This started Mr. F. looking for other 57s:

> Stranger still upon investigation is the discovery that the four are spillovers from extensive coincidences of 57 earlier in American History.... Against the laws of probability history and arithmetic came together making possible precise computations which until now were hidden in the record.
>
> [1, p. 1–2]

He found a *lot* of 57s. To show how many, it is necessary to quote at length. The "tea chest number" referred to below is 342 (6×57), the number of chests of tea that Mr. F. says were thrown into the harbor at the Boston Tea Party. When Mr. F. refers to an "alphabetical sequence number" he is using gematria to convert a word to a number using the standard $a = 1, b = 2, \ldots, z = 26$ alphabet.

> A few weeks after Lafayette's 57th birthday in September, 1814, Francis Scott Key wrote the Star Spangled Banner. When Lafayette died the Declaration of Independence was 57 years old. The alphabetical sequence numbers in the names George Washington-Thomas Jefferson-Lafayette total 456 (8×57), the sum of the double United States of America in the Declaration of Independence. The combined 16 times 57 matches the 912 days from Lexington to Saratoga. Most sources place the number of prisoners taken at Saratoga at some 57-hundred.
>
> Dec. 25, 1777, Christmas at Valley Forge: Washington's army was ragged, starving and freezing. On Christmas day, 1783, Washington was to arrive at Mount Vernon, home after victory in war. The time between the two extremes of fortune was to be exactly 57 months plus 57 weeks plus 57 days. The alphabetical sequence total of Mount Vernon is 171 (3×57).
>
> On Feb. 6, 1776, exactly 57 weeks after the Jan. 3, 1777 Battle of Princeton, France openly joined the struggle for Independence. Princeton and Yorktown, Washington's two victories over Cornwallis, were 57 months apart. The alphabetical sequence total of Princeton is 114 (2×57).
>
> The 57th month of the Revolution ended on Jan. 19, 1780, and 114 (2×57) days later, on May 12, Charleston fell to the British. Its fall was 684 (12×57) days after the Battle of Monmouth, June 28, 1776.
>
> The fateful year of Yorktown began with two battles 57 days apart, at Cowpens, N. C., Jan. 17, 1781, and at Guilford Court House, N. C., March 15.
>
> April 19, 1781, the last wartime anniversary of Lexington and Concord was exactly 57 months plus 57 weeks and 57 days after those first battles.
>
> In June, 1781, 57 Americans died in the assault on the British Fort Ninety Six in South Carolina. Ninety Six is the total for the signers of the Declaration of Independence, 57, and the Constitution, 39.
>
> When French naval forces turned back the 19 British warships sent to aid Cornwallis he became trapped at Yorktown and he surrendered on Oct. 19, virtually ending the military phase of the war that began April 19, 1775. The three 19's total 57.

> The name Yorktown fitted in exactly with what went before. The alphabetical sequence numbers of Continental Congress-Lexington-Saratoga-Yorktown total 570 (10 × 57). Multiplied by a hundred the result is 57,000, the hours from Lexington to Yorktown consisting of 3,420,000 minutes, ten thousand times the six times 57 tea chest number.
>
> After his surrender Cornwallis spent another 57 days on American soil, sailing for England with Benedict Arnold on Dec. 15, the day George III wrote a directive dated 57 days after Yorktown exhorting Parliament to continue the war.
>
> The first Christmas after Yorktown was 342 (6 × 57) days after the American victory at Cowpens which began the year. The first Fourth of July after Yorktown was exactly 57 months plus 57 weeks and 57 days after July 4, 1776.
>
> Exactly 684 (12 × 57) days after Yorktown the Treaty of Paris was signed on Sept. 3, 1783. England thereby renounced her claim to sovereignty over the Colonies 285 (5 × 57) years after John and Sebastian Cabot, in 1498, had sailed along what was to become the Colonial coast. The combined alphabetical sequence numbers of England-United States of America total 285, matching the five times 57 years.
>
> After the treaty the coincidence number 57 and the independence number 76 combined to measure the 133 days to its ratification on Jan. 14, 1784, 3,192 (56 × 57) days after Lexington and Concord. There are 3,192 hours in 133 days.
>
> Nine states were needed to ratify the Constitution. The 57 yes votes of New Hampshire overrode the nays to make it the ninth. All Constitutional law begins with the 57th word of the Constitution, that 57th word, All, being the first word of the first section of the first article. After the original seven articles the closing paragraph together with Washington's signature and titles comprise 57 words.
>
> In 1789 the young nation established the United States Government then proceeded on a 57-year march of expansion across the continent to the 1846 conquest of California.
>
> On Feb. 22, 1846, the Liberty Bell rang in its tower for the last time in tribute to George Washington, 57 years after his 57th birthday when he was president-elect.
>
> This version of history began with the four 57's listed in a column on a chart. They total 228 and the four presidents' order of service numbers in the first column 1, 3, 4 and 6, total 14. Fourteen times 228 equals 3,192 (56 × 57), the days of the Revolution. [1, pp. 18–24]

There are several possible reactions to that recital of 57s. One is boredom, since it is well known that not everyone finds everything interesting. Another is amazement, which can take two forms. The first is amazement at the number of 57s in and around the American Revolution. There are too many for them to be there by chance. Someone put them there! Who could it be? And for what reason? There *has* to be a reason! This line of thought should be resisted since the 57s are no more amazing than my possession of dollar bill number F 84323030 C. They may seem more amazing, because the number of 57s that the author has managed to quarry out of the history books is larger than the number of letters and digits in the serial number, but the same principle applies in both instances. Just as the serial number was printed on the bill before I got it, and is thus in no way amazing, the 57s all existed before Mr. F. started to look for them.

The second kind of amazement is more appropriate: amazement at the diligence and cleverness of Mr. F. in finding them all. In some places the cleverness is strained, and it shows—"Yorktown" does not gematrize to a multiple of 57, nor does "Lexington-Yorktown," but throwing in, as Mr. F. did, "Saratoga" and "Continental Congress" makes it come out. Also, there is nothing peculiar to the American Revolution that makes six years the same as 57 months, 57 weeks, and 57 days. Nevertheless, Mr. F. did amazing work. You or I would never have noticed that 96, as in Ninety-Six, South Carolina, is the sum of the number of signers of the Declaration of Independence and the Constitution.

What you should *not* feel is that there must be something behind all the 57s. All that is behind the 57s is Mr. F., finding 57s. Given as rich a field as the American Revolution, they are not hard to find. To show you how easily 57s can be had, I have found one in a much less rich field, the serial number of my dollar bill. Since F is the sixth letter of the alphabet and C is the third, we can reduce the serial number to an integer, 6843230303. Written as a product of primes,

$$6843230303 = 7 \cdot 13 \cdot 13 \cdot 17 \cdot 227 \cdot 1499.$$

If you sum the digits of the factors, you will find that the total is 57. What else could it be, on a bill with George Washington's picture on it? Speaking of George Washington, note that the sum of the last two primes in the factorization is $227 + 1499 = 1726$ and when we add 6, the number of primes in the factorization, we get 1732, Washington's year of birth. Further, 1726 is significant since $1 + 7(2 + 6) = 57$.

You may have the unworthy suspicion that I made up that serial number on purpose, just so I could get those things out of it, but I did not. The bill came directly from my wallet. The first 57 that I extracted from it did not come immediately. Finding it took effort, effort that does not show in the statement of the result. Readers of the amazing results that number-searchers find may not take this into account. I had to look at quite a few of the infinite number of ways of combining the digits in the serial number to get a 57, or a multiple of 57, which would have been almost as good. My glow of pleasure at producing the 57 that I did—I'm quite proud of it, actually—may partially explain why Mr. F. and others like him devote the enormous amount of time that they do in trying to discover 57s, multiples of 11, or whatever it is that they seek in their pyramid, their henge, or their revolution. It feels so good when you succeed.

It is to Mr. F.'s credit that he did not conclude that the 57s in the Revolution were put there by UFO space aliens or anything similar. He presented them, and let readers draw their own conclusions. He may have had space aliens on his mind, but he did not write about them.

Similarly, the 57 in my dollar bill was not put there by anyone, any more than anyone put all those 57s in the American Revolution. It did not get there, and they did not get there, from some higher intelligence, or from anyone's knowledge of ancient and lost wisdom. No conspiracy operated, darkly pulling 57 strings. Chance was at work. Just chance. Chance, that's *all*.

Reference

1. Finnessey, Arthur, *History Computed*, Atlanta, 1983.

CHAPTER 11

The Law of Round Numbers

The Law of Small Numbers has an immediate corollary:

COROLLARY (The Law of Round Numbers). There are not enough round numbers to carry out the many tasks assigned to them.

Proof. Round numbers are as rare as small numbers.

The Law of Round Numbers (LRN) is insufficiently appreciated by those who do not deal enough with numbers, small, round, or otherwise. That those who do not deal with numbers have never heard of LRN is no excuse. Those who do not deal with numbers should not use numbers to try to prove nonnumerical things.

Here follows an example of how ignorance of LRN can lead an eminent scholar to give incorrect reasons for a conclusion that he wants to reach. The conclusion may be correct (though I doubt it), but the numbers do not lead to it.

Joseph Campbell (1904–1987), the author of the four volumes of *The Masks of God* (1959–67) and many other books, was a professor of literature at Sarah Lawrence College from 1934 to 1972. His specialty was myth, about which he knew a great deal. Since no one can know everything, I suspect that his knowledge of numbers was on a par with my knowledge of the role of myth in the human imagination.

Nevertheless, he could invoke numbers if he thought that they would serve his purpose. Here is a passage from [2, p. 72]:

> One explanation that has been proposed to account for the appearance of homologous structures and often even identical motifs in the myths and rites of widely separate cultures is psychological: namely, to cite a formula of James G. Frazer in *The Golden Bough*, that such occurrences are most likely "the effect of similar causes acting alike on the similar constitution of the human mind in different countries and under different skies."

> There are, however, instances that cannot be accounted for in this way, and they suggest the need for another interpretation: for example, in India the number of years assigned to an eon is 4,320,000; whereas in the Icelandic *Poetic Edda* it is declared that in Othin's warrior hall, Valhall, there are 540 doors, through each of which, on the "day of the war of the wolf," 800 battle-ready warriors will pass to engage the antigods in combat. But 540 times 800 equals 432,000!
>
> Moreover, a Chaldean priest, Berossos, writing in Greek ca. 289 B.C. reported that according to Mesopotamian belief 432,000 years elapsed between the crowning of the first earthly king and the coming of the deluge.
>
> No one, I should think, would wish to argue that these figures could have arisen independently in India, Iceland, and Babylon.

Wrong. I would wish to so argue. They did arise independently. The LRN is operating here. Professor C. thought, consciously or not—probably not—that it is very unlikely that the same large number should appear in two or more places at once independently. Untutored intuition says that the chance that 432,000 should appear in both Iceland and India ought to be about one in 432,000. That is, very small. Untutored intuition, which also tells us that the world is flat, is in this case incorrect. The chance that the same large *round* number should pop into two different heads independently is much larger, and the chance that the same round number should be found, somewhere, in the writings of two different cultures is, I should think, very close to 1. The LRN guarantees it. There just aren't enough round numbers to go around.

It is very difficult to prove a negative, so it is conceivable that in ancient times there were couriers traveling around the world from Iceland to India, bearing with them news of 432,000, but I think it is much more likely that the figures did indeed arise independently. Frazer was wise, and Frazer had it right: similar causes produce similar effects, with no need to postulate mysterious communication.

Professor C. was a professor of literature and so could not have been expected to know the LRN, especially since it has just had its first appearance in print. Ignorance, however, is no excuse. He should have consulted someone as wise in the ways of numbers as he was wise in the ways of myth. Had he done so, he might have been asked if he would have been astonished to find that three different people in three different societies had attached to three different things the numbers 1,000,000, 100,000, and 100,000. He probably would have admitted that he would not have found that at all amazing. Those powers of ten are round numbers that naturally arise when there is a multitude to be described. Then he should be even less surprised at the appearances of 4,320,000, 432,000, and 432,000 because those numbers are even *rounder* than 1,000,000, 100,000, and 100,000.

A number is round when it is the product of many small factors. Small integers are natural and come up all the time, as the Law of Small Numbers so powerfully reminds us. Multiplication is also natural (or at least often necessary), and so products of small integers are also natural. The 60 that the ancient Babylonians used as the base of their system of numerals was very natural: it is $2 \cdot 3 \cdot 5$ with another 2 thrown in to get divisibility by 4.

We can give a quantitative measure of the roundness of an integer n. Let it be $R(n) = d(n)/\ln n$, where $d(n)$ is the number of divisors of n. So, for example, the roundnesses of the first few powers of 10 are

n	Divisors	$R(n)$
10	$1, 2, 5, 10$	$4/2.303\ldots = 1.737\ldots$
100	$1, 2, 4, 5, 10, 20, 25, 50, 100$	$9/4.605\ldots = 1.954\ldots$
1,000	$1, 2, 4, \ldots, 1000$ (16 divisors)	$16/6.907\ldots = 2.316\ldots$
10,000	$1, 2, 4, \ldots, 10000$ (25 divisors)	$25/9.210\ldots = 2.714\ldots$
100,000	(36 divisors)	$36/11.512\ldots = 3.126\ldots$

So, the more zeros a power of 10 has the rounder it is, which is as it should be.

(I did not count the number of divisors of 100000 on my fingers, or even write them down: instead I used a formula, well known to those who know it, for the number of divisors of an integer n, namely if

$$n = p_1^{e_1} p_2^{e_2} \cdots p_k^{e_k}$$

is how n is written as a product of prime powers, then

$$d(n) = (e_1 + 1)(e_2 + 1) \cdots (e_k + 1).$$

Thus, $d(100000) = d(2^5 5^5) = (5 + 1)(5 + 1) = 36$.)

Most integers are not as round as powers of 10. For example, my random-number generator gave me the following four four-digit integers:

n	$d(n)$	$R(n)$
$5316 = 2^2 \cdot 3 \cdot 443$	$3 \cdot 2 \cdot 2$	$12/\ln(5316) = 1.398\ldots$
$4525 = 5^2 \cdot 181$	$3 \cdot 2$	$6 \ln(4525) = .712\ldots$
$2112 = 2^6 \cdot 3 \cdot 11$	$7 \cdot 2 \cdot 2$	$28/\ln(2112) = 3.657\ldots$
$7790 = 2 \cdot 5 \cdot 19 \cdot 41$	$2 \cdot 2 \cdot 2 \cdot 2$	$16/\ln(7790) = 1.785\ldots.$

As you can see, three of them are less round than 10000 by quite a bit. By chance, 2112, which is rounder than even 100,000, was one of the integers generated. I could have cheated and ignored it, but I did not. Mathematicians are, in general, honest. 2112 is a *very* round number. It may not look round since it does not end with a bunch of zeros, but it is round nevertheless. I would not be at all surprised if it appears in any number of ancient documents. But

I would be even less surprised to see 43,200, because it is even rounder than 2112:

$$43{,}200 = 2^6 \cdot 3^3 \cdot 5^2$$
$$d(43200) = 7 \cdot 4 \cdot 3 = 84,$$
$$R(n) = 84/\ln(43200) = 7.869\ldots.$$

I do not know how many numbers appear in ancient documents, but I do not doubt that the number of numbers is large. When you have a large number of numbers, you have a large chance that two or more of them will be the same, especially if they are round numbers. You only need to assemble twenty-three people to have the chance that two of them have the same birthday to exceed one-half. So, if you have a collection of twenty-three numbers taken from a universe of 365 numbers, you have a better than even chance that two will be the same. Coincidences are not rare.

Professor C.'s occurrences of 432 would have been more impressive by far if he had specified in advance that 432 was the number he was looking for. If you want to assemble a group of people large enough so as to have better than an even chance that two of them have birthdays on November 6th, a date you have selected before you start gathering the group, you need to gather 253 people. That is more than ten times as many as the 23 needed for a coincidence on any day. (The reason for this is that the chance that none of n people was born on November 6th is $(364/365)^n$ and to make this less than .5 you need to take n greater than 252.65.)

It would also be impressive to have occurrences of an un-round number. If you were to show me an Icelandic poem in which there was a reference to 74,567 ice gnomes, an Indian document giving the 74,567 manifestations of some god or other, and an Egyptian papyrus mentioning the 74,567 prisoners taken in some splendid victory, *then* I would be impressed indeed. I might start to believe in the existence of prehistoric radio or some other means of communication that diffused that number from one society to another. The reason is that 74567 is a prime and primes are the least round of numbers. Round numbers are round because they are products of many small factors and primes have very few factors. The roundness of 74567 is a mere $R(74567) = 2/\ln(74567) = .178\ldots$ and it is very unlikely that three different authors would hit on that number independently.

43,200 is different. It has no prime factors greater than 5. It is natural to multiply together 2s, 3s, and 5s. There are 360 degrees in a circle because the ancient Babylonians multiplied the base of their number system, 60, by 6. We could very well have circles with 432 degrees—one sixteenth of a circle would then be the round 27 degrees—and that we do not is partly a historical accident.

The point is, round numbers are not *surprising*. There are not all that many of them, so they keep coming up in different places. That is part of what the LRN says. To have three occurrences of a very round number in the ancient literatures of three cultures is not surprising at all. Notice that Professor C. did not find three 432,000s: one was the product of 800 and 540, a multiplication that the ancient writer did not carry out.

Professor C. was trying to convince his readers that culture diffuses from a central point. That is, that new ideas are very rare and tend to be had only once. The Egyptians built pyramids and the Mayans built pyramids, so diffusionists conclude that it is much more likely that the idea of pyramids was communicated from Egypt to Central America than that it occurred to a bright Egyptian and an inventive Mayan separately and independently. This can be, and is, argued about, but I think that it is best to leave numbers out of it. Especially when, as numbers can, they take you too far.

Here is another example of Professor C. being taken too far, this time by the precession of the equinoxes.

> This almost imperceptible slippage at the slow but steady rate of one degree in 72 years is what is known as the "precession of the equinoxes." To complete one cycle of the zodiac or, as it is called, one "Great" or "Platonic Year"—requires 25,920 years; which sum, divided by 60, yields, once again, the number 432. So that the mythological count of 432,000 years not only cannot have been the product of any psychological archetype or elementary idea, but must have been discovered only through centuries of controlled astronomical observation. [2, p. 74]

This is pure numerology—that is, using properties of numbers, pure properties of pure numbers, allegedly to get information about nonnumerical reality. A Greek writer in 289 B.C. reports a Babylonian myth that has 432,000 years in it; Professor C. does some dubious arithmetic (as we will see in a minute), finds the number 432, and concludes that the 432,000 *must* have come from centuries of astronomical observation. The numbers determine history. Numbers have power over things.

Such is the power of numerology that it can suck in even highly respected scholars as Professor C. It is almost enough to make me want to have a law passed restricting the use of numbers to those with the appropriate license.

I know where the 432,000 came from:

$$432{,}000 = 2 \cdot (60)^3,$$

so the Babylonian writer had to make only *two marks* to express that gigantic span of time. In base 60, 432,000 is 2000 and since the Babylonians had no symbol for zero, it would be written as two unit marks, II. The lack of a zero meant that

$$2, \quad 2 \cdot 60, \quad 2 \cdot (60)^2, \quad 2 \cdot (60)^3, \ldots$$

would all be written the same way—two unit marks—and context would serve to tell which integer was meant. The differences between 2, 120, 7200, 432,000,... are large enough to make that possible.

Numbers in base 60 do not come much rounder than 432,000. A question is, why not use 216,000, or $(60)^3$, which could be written with only one mark? Maybe the author thought that number would be *too* round and readers would think that it was just being made up. It is the same with us: "a million" can mean nothing more than "a lot", but "2,000,000" has more of a ring of definiteness.

The number that Professor C. chose to use for the period of the precession of the equinoxes is very round:

$$25920 = 72 \cdot 360 = 2^6 \cdot 3^4 \cdot 5,$$

($R(25920) = 6.887\ldots$), so it is not surprising that it has the round 324 as a factor. There are two troubles with this integer. Let me rephrase that, since integers do not have *troubles*—numerology must be getting to me. There are two troubles with Prof. C.'s *use* of this integer.

One is that it is not right. The period of the precession of the equinoxes is not an integer, but is, to eight significant figures, 25770.036 years. This can be derived from the figure for the precession to be found in [1, p. K6] where it is given as 5029.0966″ of arc per century. If we round the period off to 25,770 years, there is no way to get 432 out of it. It is not round enough. It's pretty round,

$$25770 = 2 \cdot 3 \cdot 5 \cdot 857, \quad R(25770) = 1.575\ldots,$$

but not round enough to have a 432 in it.

Professor C. might get around that by saying that 25920 years is what the ancients *thought* was the period of the precession, but there is trouble with that as well, mainly that the precession was first noticed by Hipparchus around 125 B.C., a good century and a half after the Greek mention of the Babylonian 432,000 years.

For centuries after Hipparchus there was no agreement about what the precession was, much less its period. Some writers said that the stars go one way for a while and then reverse and go the other way. To think that pre-Hipparchian Babylonians not only knew about the precession but had estimated its period to within one percent is absurd. Of course, there is always the rebuttal that the records of this ancient wisdom have been lost, as was the ancient wisdom, but then we have descended to the level of explaining events by invoking the

Atlanteans or alien visitors, and that far I do not think that Professor C. would have wanted to go.

What happened was that an unknown writer, knowing nothing at all about the precession of the equinoxes, picked the round number 432,000 out of the air because there was need for a big number. Another unknown writer picked the round number 4,320,000 out of the air because of a similar need. A third happened to use 540 and 800, both round numbers, whose product is the rounder number 432,000. There is no cause for amazement, since *thousands* of unknown writers have had occasion to pick big round numbers out of the air because they needed one, and that two would hit on the same one (or nearly the same one—only an order of magnitude different) is not surprising, but inevitable. The Law of Round Numbers guarantees it.

The lesson to be learned is that the impulse toward numerology is not restricted to the ignorant and the eccentric. As Professor C. shows, it can exist even in keen and highly trained brains. That implies that numerology may lurk, to some degree, within us all. It is part of how we are made. We have to watch out, or it may get us.

References

1. *The Astronomical Almanac*, Science and Engineering Research Council, Washington and London, 1994.
2. Campbell, Joseph, *The Mythic Image*, Princeton University Press, Princeton, 1974.

CHAPTER 12
Biblical Sevens

The title page of *Mathematics Prove Holy Scriptures* [1] has, below the title and the demonstration that the number of Jesus is 888 and of Satan is 666, the following statement:

> Within these pages—before your very eyes—is an actual scientific demonstration of the Scriptures!

It is curious that so many fundamentalist Christians want to make the Bible *scientific*. Evidently it does not suffice to say, "Faith tells me that the universe and all that is in it was created in 4004 B.C." There must be scientific proof. Around Christmas every year you can find, without even looking very hard, articles scientifically explaining the star of Bethlehem and, if you look at more than one of them, you will have your choice of more than one scientific explanation. I am no expert on such matters, but these efforts and others like them seem to me to diminish the status of an all-powerful deity. If you are omniscient and omnipotent, then if you want to put a star in the sky you just *do* it. You don't need any excuse, or any help.

Be that as it may, books such as this one continue to appear. Probably their authors hope that those who value and respect the conclusions of science will read them and therefore be convinced. Their authors cannot lose: if even one mind is changed they have a success, and it is not likely that their efforts will unconvince any of the already convinced. Thus, unless you are anti-Christian, you may think that such efforts can do no harm and may do some good.

But it is not good when, as in this book, the claim is made that mathematics *proves* the divine inspiration of the Bible. Mathematics does nothing of the kind. That is not the business of mathematics. Numbers have power, but not that sort. Mathematics is being misused when it is pressed into service to prove divine inspiration. Misuse proves nothing at all and this book and others like it are therefore to be deplored. We must resist the spread of error, firmly and always.

The book is based on the life's work of Ivan Panin, who was born in Russia on December 12, 1855 and died in Canada on October 30, 1942. The author gives a brief biography:

> As a young man he was an active Nihilist and participated in plots against the Czar and his government. At an early age he was exiled from Russia. After spending a number of years in Germany furthering his education, he came to the United States. Soon after his arrival, Mr. Panin entered Harvard University. He was a personal friend of the famous Professor William James, and President Eliot of Harvard.
>
> Mr. Panin is a brilliant scholar and a Master of Literary Criticism. After his college days he became an outstanding lecturer on the subject of literary criticism. He lectured on Carlyle, Emerson, Tolstoy, and on Russian literature, etc., being paid as high as $200.00 for each address.... During this time Mr. Panin became well known as a firm agnostic—so well known that when he discarded his agnosticism and accepted the Christian faith the newspapers carried headlines telling of his conversion. Professor James, who was reputed to be the greatest Metaphysician of his time, remarked, "What a pity that Mr. Panin is cracked on religion. A great philosopher was spoiled in him." Prior to his conversion Mr. Panin wrote some three thousand aphorisms and many remarkable essays, which are indeed a memorial to his days as a Master of Literary Criticism. Mr. Panin was also the editor of two daily newspapers. He is a gifted writer and a brilliant and eloquent speaker.
>
> [1, pp. 110–111]

Here is how he got cracked on religion. In 1890,

> Mr. Panin was casually reading the first verse of the Gospel of John in the Greek—"In the beginning was the Word, and the Word was with (the) God, and the Word was God." The question came to his mind, "Why does the Greek word for 'the' precede the word 'God' in one case, but not in the other?"
>
> Therefore in one column he made a list of all the New Testament passages in which the word "God" occurs **with** the article "the," and in another column he made a list of all the passages in which the word "God" occurs without the article. On comparing the two sums he was struck with the numeric relation between them. He then followed the same procedure on the word "Christ" and on other words, and found amazing numeric facts. This was the beginning of the profound numerical discoveries which are now called the Science of Bible Numerics.
>
> Since discovering that first feature in 1890 Mr. Panin has earnestly devoted his entire life to one definite and specific purpose. For the past half century he has toiled faithfully and tirelessly with undaunted persistence. He has devoted himself so persistently to counting letters and words, figuring numeric values, making concordances, and working out mathematical problems, that on several occasions his health completely failed. Regardless of the tremendous mental and physical strain he has labored faithfully and diligently for the past fifty years. [1, pp. 112–113]

Fifty years!

> The original manuscripts of his work consist of approximately 40,000 pages. Upon these pages, many of which are now yellow with age, Mr. Panin has carefully and thoughtfully written millions of figures. The completion of his work has required persistent labor from 12 to 18 hours daily throughout the past fifty years. [1, pp. 109–110]

Forty thousand pages! Twelve hours a day, every day! For fifty years!

For what? For finding sevens in the Bible. Not the many sevens that appear explicitly, but others:

> The sevens are strangely out of sight of ordinary Hebrew and Greek readers. They are mysteriously hidden. Thousands who have read and studied the original Hebrew and Greek text of the Bible have passed by these strange occurrences of the number seven without even noticing or knowing of their presence. These sevens are so deeply concealed that special searching and investigation and special counting are necessary in order to find them. They are said to occur "beneath the surface" or "in the structure" of the text because they are beyond the observation and view of ordinary Hebrew and Greek readers.
>
> It has been discovered that hundreds of these sevens are mysteriously hidden in the structure of the text. Some of the sevens are strangely concealed in the unusual system of numbers—in the "numeric values" of the Hebrew and Greek letters, words, sentences, paragraphs and passages of the text, while other sevens are hidden in other remarkable and peculiar ways.
>
> These sevens beneath the surface of the original Hebrew and Greek Bible text are the amazing newly discovered facts. The astounding new discoveries are numerical discoveries—discoveries having to do with numbers. The recently revealed numerical facts enable us to see before our very eyes an actual scientific demonstration of the divine verbal inspiration of the Bible. [1, p. 18–19]

Why sevens? I suppose because of the number-mystical associations of the digit. One indication of King Lear's madness was his failure to appreciate them (*King Lear*, Act 1, Scene 5):

> *Fool.* The reason why the seven stars are no more than seven is a pretty reason.
>
> *Lear.* Because they are not eight?
>
> *Fool.* Yes, indeed.

The sevens were discovered by gematria, using the usual Greek and Hebrew alphabetic system with no variations, and by counting. So, as usual, the number of Jesus is 888:

$$\text{Ιησους}, \quad 10 + 8 + 200 + 70 + 400 + 200 = 888.$$

The answer to the natural question of why Jesus is gematrized in Greek when he certainly did not write his name in that language is no doubt answered by saying that all of the earliest New Testament manuscripts are in Greek. But when we consider the Old Testament, we must use Hebrew. That is how Mr. S. warms up, with the first verse of the first chapter of Genesis:

בדאשיח	913	In the beginning
כדא	203	Created
אלהים	86	God
אח	401	(An indefinite article which is not translatable)
השמים	395	The heavens
ןאת	407	And (with indefinite article)
האדץ	296	The earth

Now come the sevens. The sum of the values of "God," "heavens," and "earth" is

$$86 + 395 + 296 = 777.$$

a multiple of 7.

The total numeric value of the three words, strange as it may seem, is exactly 777, which is of course 111 7's.

Is it not strange that the numeric value of these words is a value which divides perfectly by 7—a value which is an exact multiple of seven? Notice that the numeric value of the words is not 776 or 778, but exactly 777. If the numeric value were 776 or 778 it would not divide evenly by 7.

In addition,

created—203—is a multiple of 7,
the sum of the first, middle, and last letters is $133 = 19 \times 7$,
the sum of the first and last letters in all seven words is $1393 = 199 \times 7$,
the sum of the first and last letters in the first and last word is $497 = 71 \times 7$,
the sum of the last letters in the first and last words is $490 = 70 \times 7$,
the number of words in the verse is 7,
the total number of letters in the verse is $28 = 4 \times 7$,
the number of letters in the first three words is $14 = 2 \times 7$,
the number of letters in the last two words is 7,
the number of letters in the third, fifth, and seventh words is $14 = 2 \times 7$,
the number of letters in the third and fourth words is 7, and
the number of letters in the fourth and fifth words is 7.

That certainly is a lot of sevens. But do they *prove* anything? Here is a similar example, originally written in English so that it can be gematrized by the $a = 1, b = 2, \ldots, z = 26$ system. The sevens in it were not found by Mr. S., but by me. The time it took was approximately 28 ($= 4 \times 7$) minutes.

Lincoln's Gettysburg Address
November 19, 1863
Fourscore and seven years ago our fathers brought forth on this continent a new nation, conceived in liberty, and dedicated to the proposition that all men are created equal.

"Lincoln" has seven letters, as does "address."

The value of "Gettysburg," added to its number of letters, is $154 = 22 \times 7$.

"Liberty" has seven letters and its value is $91 = 13 \times 7$.

"All men," "created," and "equal" have equal values, all $56 = 8 \times 7$! The chance of having three consecutive 56s is approximately $1/56^3 = .000005674$ and the chance that they will all be divisible by 7 is one-seventh of that, or .000000813. This is far smaller than the value of .05 that statisticians use to establish statistical significance and thus shows that the multiples of seven cannot be accidental and must have been planned.

"Fathers" and "brought" both have seven letters and both have values that are multiples of seven, 7×11 and 7×13.

The date of the address, 11-19 1863, has value, interpreting the dash as a minus sign, of $1855 = 265 \times 7$.

The entire text, counting the title and date, contains $35 = 7 \times 5$ words.

The value of "forth on this continent" is a multiple of seven ($266 = 38 \times 7$), as is that of "a new nation conceived" ($196 = 4 \times 7 \times 7$).

I am sure that there are more sevens that could be extracted from the passage ("all men are created equal" has $21 = 3 \times 7$ letters) but the point, that any text can be made to yield sevens, I hope has been made.

It is impossible to prove that Lincoln did not plan all of the sevens, or to prove that, if he did not plan them, a higher intelligence inspired him to write them in. But, if *any* text can be made to yield sevens in abundance, are we to conclude that they were *all* planned, or *all* inspired? I do not think so. This paragraph may contain numerous sevens, but none was planned ("planned"—seven letters!), nor did I feel at all inspired when I wrote it.

By the way, I hope that no indignant reader will point out that since 56 is divisible by 7, dividing .000005674 by 7 to get .000000813 is absurd. Of course it is, but numerologists pull similar tricks all the time.

Returning to Mr. S., here is a New Testament example, his list of what he calls "features" to be found in the Gospel of Matthew, Chapter 1, verses 1–17. This is the account of the genealogy of Jesus. Mr. S., or Mr. Panin, asserts that the passage contains 72 vocabulary words. That is, repetitions and different forms of the same word do not count as different words. Here, given considerably more briefly than in the book, are the nineteen features of the passage. I have not checked any of them. Mr. Panin gematrized from the original Greek, or as close as he could get to it.

1. The total numerical value of the 72 words is 42,364, a multiple of 7.
2. The 72 words occur in 90 forms, whose total is 54,075, another multiple of 7.
3. Of the 72, 56 are nouns.
4. "The" occurs 56 times.
5. The number of different forms of "the" is 7.
6. The number of vocabulary words used in the first 11 verses is 49.
7. Of these 49, the number that begin with a vowel is 28.
8. The number of letters in the 49 words is 266.
9. Of these 266 letters, the number of vowels is 140.
10. Of the 49 words, the number that occur more than once is 35.
11. Of the 49, the number that appear in only one form is 42.
12. Of the 49, the number of nouns is 42.
13. Of the 42 nouns among the 49 vocabulary words in the first 11 verses, 35 are proper names.
14. The number of letters in the remaining 7 words is 49.
15. The 35 proper names occur 63 times.
16. Of the 35 proper names, 28 are male.
17. The 28 male names occur 56 times.
18. Three women are referred to in the first eleven verses: the number of letters in their names is 14.
19. Babylon is the only city referred to. Its name has 7 letters.

[1, pp. 27–33]

That is an impressive number of sevens. In fact, I suspect that it is *all* of the sevens that could be found in the passage with fifty years of looking (if he could have squeezed out two more, the total number of features would have been $21 = 3 \times 7$), but that is by the way. What is not by the way is the argument that Mr. S. makes next. He asks:

> How can one account for the presence of these facts?
>
> How did these features occur or come to be in the very structure of the Bible text? Their presence must be explained in some satisfactory way.
>
> The reader will agree that there are only two possible ways in which these amazing facts could have occurred. They could have occurred either **by accident**, that is, by sheer chance, or **by design**. [1, p. 50]

Whenever a writer slightingly refers to "sheer" chance, you can be sure that chance is going to lose out to design. Sure enough, there follows the argument that you would expect:

> What are the chances for [the features] to occur accidentally?
>
> According to the law of chances, for any 1 number to be a multiple of 7 accidentally, there is only one chance in 7.

> According to the law of chances, for any 2 numbers to be multiples of 7 accidentally, there is only one chance in 7 × 7, or only one chance in 49.
> [1, pp. 50–53]

And up we go. There is a table, meant I think to be impressive, in part including

For 12 features* one chance in	13,841,287,201
For 13 features* one chance in	96,889,010,407
For 14 features* one chance in	678,223,072,849
For 15 features* one chance in	4,747,561,509,943
For 16 features* one chance in	33,232,930,569,601
For 17 features* one chance in	232,630,513,987,207
For 18 features* one chance in	1,628,413,597,910,449
For 19 features* one chance in	11,398,895,185,373,143

* "to occur accidentally, there is only"

It is clear what this is leading up to.

> When there is only one chance in thousands for something to happen accidentally, it is already considered highly improbable that it will occur at all. When there is only one chance in hundreds of thousands, it is considered practically impossible. But here there is one chance in not only millions, but billions, and trillions...
>
> The amazing numerical features of even one small passage, to say nothing of the thousands in the entire Bible, could not possibly have occurred by accident—by sheer chance. All this evidence simply cannot be explained by the doctrine of chances. If those features did not occur accidentally, then there is only one conclusion, one alternative. They were purposely designed or arranged—their occurrence in such a marvelous and mysterious manner was intended or planned. [1, pp. 53–54]

The question is, was Mr. S. being ignorant or dishonest when he made this argument? If he were to put the integers from 1 to 7,777,777,777 into a hat and invite me to draw 19 of them (with or without replacement, it makes hardly any difference), then there would indeed be only one chance in billions and trillions that I would draw nothing but multiples of seven, and if that event occurred I would be safe in concluding that something other than chance had been operating. If he were to give me the opportunity of specifying my own nineteen numbers, things of which I had no advance knowledge, such as

> the number of nouns in the book of Amos,
> the sum of the first and last letters of the words in II Kings 2, 23–25,
> the sum of all the numbers in the book of Numbers,
> the number of times "all is emptiness" appears in Ecclesiastes,

and so on for fifteen more, and all those nineteen numbers were multiples of seven, then I would be amazed. Actually, I would be utterly stunned, since the chance that all are multiples of 7 is $1/7^{19}$, or about 9×10^{-17}. The chance that more than seven of the numbers are multiples of seven is already negligible. If all nineteen were multiples of seven, I don't know *what* I would do.

Picking features out of something that already exists and asking what the probability is that they exist is not playing fair. The probability is 1. What is the chance that you exist? When you consider the number of possible ways of combining genes, the probability of getting the combination that makes you the person you are is considerably smaller than $1/7^{19}$. In fact, it makes $1/7^{19}$ look like a sure thing. It is so close to zero that you are impossible, the laws of chance say so. But there you are, not only existing but, all the more improbably, reading this.

There is no difficulty in finding features in any collection of numbers, given enough of them. Here are some random digits, just as they come from my $12.95 (1295 is divisble by 7) pocket calculator:

03669 69965 71332 63411 48133 69286 69806 79351 13881

91333 70468 61886

The third through the ninth digits are all either six or nine (the chance of that happening by accident is $(2/10)^7 = 1/78125$) and their sum is $51 = 3 \cdot 17$. The second, sixth, and ninth of those five-digit groups are the only ones that are divisible by 7, and $2 + 6 + 9 = 17$. Also, the sixth group is 9898×7, and $9 + 8 = 17$. Further, $9898/7 = 1414$ (only 1 chance in 49 of *that* happening), and $17 - (1 + 4 + 1 + 4) = 7$. Also, $13881/7 = 1983$, the sum of whose digits is a multiple of 7. In addition, the sum of the first ten groups is 7×82881, and the sum of the digits of the quotient is $3 \times 3 \times 3$, echoing the threes to be found in the first ten digits. And so on. I have no doubt that the sequence contains many other amazing features that I am too lazy, or not sufficiently ingenious, to discover. But I could sum them all up by saying that the probability that the sequence occurred by accident is $1/10^{60}$, so small as to be negligible.

After establishing to his satisfaction that the numerical features could not have occurred by chance and must have been put there, the natural question that Mr. S. asked is

> Who planned and carried out these amazing numeric designs? [1, p. 55]

The answer is clear to Mr. S.:

> If anyone could write an intelligible passage of 300 words, constructing it in such a way that the same designs and schemes occur in the structure of its text as those found in the first seventeen verses of Matthew, indeed he would prove himself a wonder if he could accomplish it in six months....

> No mortal in the lifetime of a hundred years could possibly have carried out the design which is found in even a single book of the Bible, if he devoted the entire hundred years to the task. It would require some centuries to write a book designed and constructed in the manner in which the book of Matthew is constructed. [1, pp. 55–56]

Thus,

> every candid, logical minded individual is simply compelled to admit that the intelligence which planned and designed the Bible must have been Superhuman, Divine. That One Designer was a Supernatural, Master Designer. Only the Supreme, Omniscient, Omnipotent God could have caused such phenomenal numeric designs to occur beneath the surface of the Bible text. Only God could have constructed the Bible in the amazing manner in which it is constructed. The Eternal, Omnipotent Author designed, superintended, worked, and carried out His Own infinite plans. There is no escape from this conclusion. [1, p. 63]

There is also no escape from the question of why God should choose to toy with humanity by setting it puzzles simultaneously obscure, tormenting, and silly, but Mr. S. does not go into this. Mr. S. was unaware of other Bible-searchers, whose work appears in the next two chapters, who found elevens, thirteens, squares, and triangular numbers. Which did God put there? *All* of them? If so, why? God is diminished by being made into a fabricator of puzzles.

Mr. S. did anticipate the question of whether other books displayed features like the Bible. No, he said, they did not.

> Various professors of Greek have been requested to submit Greek prose classics to the numeric test to ascertain if the same amazing numerical phenomena could be found. No one has reported any success at finding such numeric designs. [1, p. 62]

I think it is a good bet that the reason that none of the professors of Greek reported success is that none of them reported anything. They did not try. Mr. Panin suffered the common fate of the crank, being ignored.

> At one time, a number of years ago, Mr. Panin sent some of his numerical material with vocabularies and other details to nine rationalists, whom he respectfully and publicly invited to refute him... saying in print, "Gentlemen, will you kindly refute my facts; will you refute the conclusions?"
>
> Mr. Panin received several replies. One was not "interested" in his "arithmetical" facts, two "regretted" that they "had no time" to give heed thereto. Another "did not mean to be unkind, but..." The others were silent....
>
> No attempt has been made to deal with Mr. Panin's undeniable facts or with the unavoidable conclusion. A destructive critic can do nothing in the way of refuting these facts—all he can do is ignore them. Are Mr.

> Panin's figures wrong? If so, where are they wrong? If they are not, then the inferences are indisputable. One cannot argue about mathematics.
>
> [1, pp. 123–124, 126]

How true, how true. The last sentence only, that is.

Reference

1. Sabiers, Karl, *Mathematics Prove Holy Scriptures*, 1941; 1969 edition published by Tell International, Los Angeles.

CHAPTER 13
Thirteens and Squares

In the last chapter, we had the deluge of Biblical sevens found by Ivan Panin. In the next, we will see thirty-sevens and triangular numbers in the first verse of Genesis. In this one, we have thirteens and squares.

The natural question is, why did God go to the trouble of inserting in the Bible so many different kinds of numbers? The natural answer is, they were not inserted at all.

In [2] we have an exercise in Hebrew gematria.

> Gen X, 21–32, deals with the genealogy of Shem. The descendants of Shem may be divided into 2 groups. The first group contains 13 names, from Elam to Joktan. The second group covers the 13 sons of Joktan. The numerical value of the names of the first group is 3588; of those in the second, 2756. Both numbers are multiples of 13.

And neither is a multiple of seven.

> Gen. XXXVI deals with the genealogy of Esau. The descendants of Esau are enumerated twice. First 15 descendants comprising sons and grandsons, then 11 descendants who later became Edomitic dukes. The sum $15 + 11 = 26$ is a multiple of 13 and the numerical value of the corresponding names adds up to $4186 = 322 \times 13$, while the numerical value of the names of the Edomitic dukes $= 3211 = 19 \times 13^2$.
>
> It is interesting that the Gematria of the names of the persons listed in the same chapter but not belonging to the above two categories also adds up to a multiple of 13, namely $4472 = 344 \times 13$.
>
> The "Gematria" of the names of the members of the immediate family of Esau is $2197 = 13^3$.

Multiples of 13 are approximately half as common as multiples of 7 and thus less likely to appear by chance. In fact, finding thirteen multiples of 13 is $(13/7)^{13} = 3126.0006\ldots$ times less likely than finding thirteen multiples of 7. And look at that number 3126! Its second half, 26, is a multiple of 13 and its first half, 31, is 13 reversed. Is that not highly significant? No, it is not: it is the Law of Small Numbers once again working its magic.

There are more thirteens in Genesis:

> Gen. XXXVI, 20–28, deals with the genealogy of Seir, the ancestor of the aborigines of the land of Edom. The number of male descendants of Esau is $26 = 2 \times 13$. The sum total of the numerical values of the male descendants, if the name of the female descendant, Timnah, is added is $512 \times 13 = 2^9 \times 13$.

The name of Timnah was included for the sole reason that it makes the sum come out to be a multiple of 13.

> The number 13 appears in the letters JHVH of the name of God: Jod = 10, He = 5, Vaw = 6, He = 5, $10 + 5 + 6 + 5 = 26 = 2 \times 13$, also in the letters of the word *Echad* (one), expressing the most important attribute of God—his oneness. The calculation of the Gematriae cannot help recalling to the mind of the reader the first principle of the Bible: God is One.

God may be One, but Bible-numerists are many: which is it that the Bible is full of, sevens or thirteens? Or could it be squares? Here is an excerpt from an advertisement [3] whose very large headline is ENOUGH OF THESE BABBLINGS, with the subhead "Enough of these babblings which say there is no evidence that supports the inspiration of the Bible." Using gematria, which the anonymous author of the advertisement chose to call "theomatics," we see that

> "God" has a value of 484, or $(22)^2$.
> "Holy" is 484, or $(22)^2$.
> "Spirit," has a value of 576 or $(24)^2$.
> God has chosen to represent these perfect things as perfect squares.
> To enumerate a few more of these, "Image of God," is $(37)^2$; "For they shall see God," is $(54)^2$, "Shepherd," is $(16)^2$, "Giving thanks," is $(45)^2$.

Following some material on 153, we get back to squares:

> God has put an additional proof in this and other texts pertaining to miracles.
> The theomatic value for "Lazarus" is 144, or $(12)^2$.
> "Come out" is 1444, or $(38)^2$
> "Came out the one having died having been bound the feet and the hands with bandages" is 3600, or $(60)^2$
> "Loose him and let him go" is 3969, or $(63)^2$.
> The odds against just these three passages being perfect squares is 200,000 to one. But that's not all. We find that whenever a miracle is performed by Jesus, a similar perfect square system is found.
> In the ninth chapter of John, Jesus heals a blind man, "When he had thus spoken, he spat upon the ground, and made clay of the spittle, and he anointed the eyes of the blind man with clay" (John 9:6). This passage has a theomatic value of 9409, or $(97)^2$

"And said unto him—" $(41)^2$. "Go, wash in the pool of Siloam," $(45)^2$.

"Pool" itself has a value of 729, or $(27)^2$.

In the fifth chapter of Mark, Jesus is called to a home where a little girl has died. "And he took the damsel by the hand and said unto her, TALITHA CUMI, which is, being interpreted, Damsel, I say unto thee, arise" (Mark 5:41).

"And straightaway the damsel arose," $(50)^2$.

"And they were astonished with a great astonishment," $(52)^2$.

And he ordered that, "No man should know it," $(23)^2$.

In the third chapter of Mark, a man with a withered hand is brought to Jesus. Jesus told him, "Stretch forth thine hand," $(35)^2$. The man stretched forth his hand, "And his hand was restored," $(31)^2$.

In the ninth chapter of Luke, a boy who was possessed of a demon was brought to Jesus. Jesus rebuked "the unclean spirit" $(46)^2$; of the "demon" $(26)^2$; and "cured the boy" $(26)^2$.

The story continues as Peter's mother-in-law arose and served them. That night many sick people were brought to the house. "And evening coming, they brought to him many who were demon possessed, and he expelled the spirits with a word, and all the ones ill he healed," $(112)^2$.

"He expelled the spirits with a word" is $(44)^2$.

"They brought to him many who were demon possessed" is $(64)^2$.

How does this prove anything about miracles? We have seen just twenty-five of the examples of perfect squares which God put into the Greek in which the Holy Spirit directed the writing of the New Testament whenever miracles were described. The probability of these perfect squares occurring by chance is computed for just these twenty-five examples.

The odds against these twenty-five examples are one in 400,000,000,000,000,000,000,000,000,000,000.

I would not want to bet against these odds.

In fact, no mathematician would. **There can be only one explanation and that is that *God purposely put* this system of perfect squares in the Bible when miracles are related so that you and I and the rest of the skeptics about miracles could have some of the solid, scientific proof we need.**

It is not clear how that number with all the zeros at its end, 4×10^{32}, was obtained. Since there are twenty-five perfect squares, probably assumed to have occurred independently, you would expect the number to be the twenty-fifth power of something. Of course it is a twenty-fifth power, but of $20.141\ldots$, a number whose significance I cannot guess at. It seems more likely that the author of the advertisement typed zeros until fatigue set in.

Wherever the number came from it is, of course, nonsense. Exactly the same argument would show that the odds against the existence of the pocket calculator I used to calculate the twenty-fifth root of $4 \cdot 10^{32}$ are 17,576,000 to 1. The reason is that it is a model EL-546G. There is only one chance in 26 that the first letter should be an E, the same chance that the second letter should be an L, one chance in ten that the digit after the L should be a 5 and so on: the chance of the calculator is one in

$$26 \times 26 \times 10 \times 10 \times 10 \times 26 = 17576000.$$

A Bible-numerist would say that this is so small that it proves that some outside influence caused me to have that calculator. Nonsense! The probability of the existence of something that exists is 1, no matter what calculations you carry out.

The advertisement ended:

> The text of this ad is comprised of excerpts from the book "A Scientific Approach To Christianity" by Robert W. Faid, (Published by New Leaf Press). Available at any Christian bookstore.

Those curious about Biblical squares and theomatics could perhaps satisfy their curiosity by looking for the book, something I have not yet done.

Let us apply this method to another text, Lincoln's "Gettysburg Address," and see if it has any squares in it. The scope for finding them is much less than in the Bible since the address is so short.

There follows a list of some of the squares that I found. Since Lincoln was writing in English, I gematrized using the natural-order alphabet, $a = 1$, $b = 2, \ldots, z = 26$:

altogether fitting	196
we have come	100
power to add	121
a new birth	100
birth of freedom	100
our fathers brought forth	289
a new nation conceived	196
the great task remaining	225
dedicated to the great task	225
civil war testing whether that nation	400
poor power to add or detract	289
and that government of the people	324
on a great battlefield of that war	289
the proposition that all men are created equal	441
nation, or any nation so conceived and so dedicated	441
forth on this continent a new nation, conceived in liberty	576
we have come to dedicate a portion of that field	400

to be here dedicated to the great task remaining before us	484
who struggled here have consecrated it, far above our poor power to add	676
unfinished work which they who fought here have thus far so nobly advanced	729
increased dedication to that cause for which they gave the last full measure of devotion	841
that these dead shall not have died in vain—that this nation, under God, shall have a new birth	841

That is only twenty-two squares, but consider the brevity of the text compared with the Bible. I calculate that the odds against twenty-two squares are about 41,943,040,000,000,000,000,000,000,000 to 1, and I would not want to bet against those odds.

But what, exactly, would I be betting against? That God wrote the Gettysburg Address? We may be sure that Abraham Lincoln, by tradition writing in his rattling and swaying railroad car, with neither a table of squares nor an electronic calculator, did not put all those squares in his text on purpose. And, those who find sevens, thirteens, and squares in the Bible would argue, so many could not possibly be there by accident.

But they *are* there by accident. No one put them there. To try to use numbers to prove something that is not about numbers is foolish. I think that I could find a few sevens, thirteens, and maybe even a thirty-seven or two, in the address if I looked. God did not put *them* there, either. By the way, "Gettysburg" gematrizes to 144, a perfect square, and God did not do that. The Law of Small Numbers did it.

The Bible-numerists, displaying the human ability to see patterns where none exist, to attribute volition where there is none, and to invest the operations of blind chance with deep meanings they do not have, go too far. Even if they all were united in finding sevens they would not be convincing, but the multiplicity of their findings make each of their hypotheses even less likely. What Augustus De Morgan said about people who square the circle applies to Bible-numerists as well:

> If I had before me a fly and an elephant, having never seen more than one such magnitude of either kind; and if the fly were to endeavor to persuade me that he was larger than the elephant, I might possibly be placed in a difficulty. The apparently little creature might use such arguments about the effect of distance, and might appeal to such laws of sight and hearing as I, if unlearned in such things, might be wholly unable to reject. But if there were a thousand flies, all buzzing, to appearance, about the great creature; and, to a fly, declaring, each one for himself, that he was bigger than the

> quadruped; and all giving different and frequently contradictory reasons; and each one despising and opposing the reasons of the others I should feel quite at my ease. I should certainly say, My little friends, the case of each of you is destroyed by the rest. [1, vol. 1, p. 1]

This is not to say that there are not numbers in the Bible. The ancient authors and scribes of the Biblical text, writing in languages in which letters were numbers and numbers were letters, probably influenced by number mysticism, and no doubt just as ingenious as the Bible-numerists, could very well have inserted on purpose some sevens and thirteens here and there. Today's writers, writing in languages in which letters are not numbers, do not think of doing such things at all, so some things that seem a little farfetched to us may in fact have been intentional. I find it hard to credit that the numerical values of the names of the Edomitic dukes came out to be a multiple of 13 because a writer made it come out that way on purpose (by changing a spelling slightly, perhaps, if the need arose), but that may be because my mind and the mind of the ancient writer are not alike.

Even so, I think the 13 was coincidental. However, it seems well within the bounds of possibility that one thing that a Bible-searcher noticed was done on purpose:

> The story of the Tablet of the Ten Commandments is written in two installments in both Exodus and in Deuteronomy. In each of the four installments the word "Tablets" occurs seven times.

If it did not already exist, that pleasing symmetry could have been created by making small adjustments, easily made if the writer wanted to. Dante arranged many things in *The Divine Comedy* by threes and nines: it pleased him to do it, and it was very little trouble. There are many sevens in the Bible and I suspect that not all of them occur because of the Law of Small Numbers. Some were included on purpose.

Putting in a seven here and there differs from the elaborate schemes of our authors in the last chapter, this one, and the next. Why such elaboration? Why such obscurity? Why such puzzles? The answer, I suppose, would be that it is not for miserable humans to question the ways of God. There is no way to refute the statement that God arranged the Bible just as the Bible-numerists say, just so they could find their sevens and their squares. Nor is there any way to refute the statement that the universe was created yesterday, including all of us and all of our memories. Appeals to plausibility are all that we have.

It would probably do no good to show one Bible-numerist the work of another. I once put two people who thought that they had trisected the angle with straightedge and compass in touch with each other, and each concluded that the other was crazy. The result with Bible-numerists would probably be similar.

Bible-numerists' faith must be shaky. If it weren't, they wouldn't find it necessary to support the Bible with their numbers.

References

1. De Morgan, Augustus, *A Budget of Paradoxes*, London 1872; reprinted with notes by D. E. Smith by Open Court, Chicago, 1915; reprinted by Dover, New York, 1954.
2. Goldberg, Oscar, On numbers in the Bible, *Scripta Mathematica* **12** (1946) 231.
3. Nashville *Tennessean*, October 6, 1992, page 7-A.

CHAPTER 14

The Triangles of Genesis 1:1

In chapter 12 we saw how Ivan Panin found many sevens in Genesis 1:1, leading him to conclude, after finding many sevens elsewhere in the Bible, that there were too many sevens to be accidental and thus they must have been put there by God.

Mr. P. missed something. He missed all the triangles, 37s, and 73s in the same verse. V. J., who has circulated his work only by correspondence (what I have seen is dated 1990–1992) has found them and has concluded that there are too many triangles, 37s, and 73s to be accidental and thus they must have been put there by God.

Who is right, Mr. P. or Mr. J.? Could it be that they are *both* right and that God put both the sevens *and* the triangles, 37s, and 73s there to be found by different people so that they could come to the same conclusion? That is, I suppose, possible, but the motive escapes me. It seems much more likely that we have two ingenious people independently concentrating on the same material and finding in it two different patterns, patterns that were not put there by anyone.

Here is Genesis 1:1 again, gematrized:

האד׳ן	ואח	השטים	אח	אלחים	בדא	בדאשיח
296	407	395	401	86	203	913

The patterns do not leap to the eye, but not only are they there, they are not too hard to find, Mr. J. says:

> Attached to the opening words of the Bible, as though by some divine superglue, is a structure of pure number that is rooted in the heart of mathematics. However, this is not to say that it is inaccessible to the average person; on the contrary, it is open to all who have an eye for symmetry, and are able to count—and most possess these modest qualifications!

Mr. J. notes first that the number of letters in the seven words of the verse is 28, the seventh triangular number, shown in Figure 1.

FIGURE 1 The seventh triangular number.

In addition, if you count the number of dots around the outside of the triangle, you get $18 = 6 + 6 + 6$, and 6 (another triangular number) and 28 are the first two perfect numbers. Further, if you interpenetrate two 28-triangles, you get a star of David with 37 dots, as in Figure 2. Yet further, the hexagon interior to the star has 19 dots (Figure 3). These numbers will be back later.

You may not think that this is very impressive, but when you add the values of the words in Genesis 1:1 the total is 2701. Mr. P. passed this over with no notice, perhaps because 2701 is not evenly divisible by 7, but it is a noteworthy number. It is the 73rd triangular number. Also,

$$2701 = 73 \times 37.$$

There those numbers are again!

Not only that, the number of dots on the periphery of a 2701-dot triangle with 73 dots on each edge is

$$72 + 72 + 72 = 216 = 6 \times 6 \times 6,$$

a perfect cube. Of the 4472 triangular numbers from 3 to 10,000,000, Mr. J. said, only seven have perimeters that are cubes.

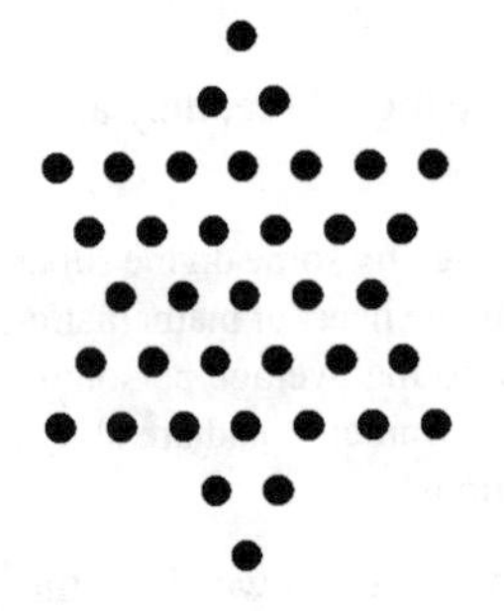

FIGURE 2 A 37-dot star of David.

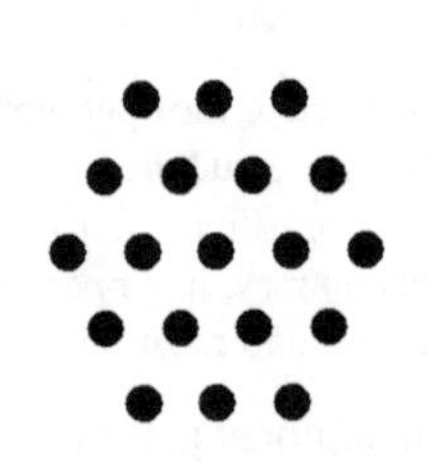

FIGURE 3 A 19-dot hexagon.

He is right, but I hope he did not examine each of the 4472 triangular numbers one at a time. The number of dots on the perimeter of the nth triangular number is $3(n-1)$. If this is a cube, then

$$3(n-1) = r^3.$$

This implies that r is divisible by 3, so $r = 3k$. Thus

$$3(n-1) = 27k^3$$

or

$$n - 1 = 9k^3.$$

Thus, the only times the nth triangular number has a perimeter with a number of dots that is a perfect cube are when $n = 1 + 9k^3$, and for $1 < n \leq 4472$ there are seven such, one being $n = 73$.

Note that if you take the 37-dot hexagon in Figure 4 and append a six-dot triangle to each of its six sides you have the 73-dot star of David in Figure 5. Notice the recurrence of 37 and 73. Then if you remove the 36 perimeter dots from the star, you have the 37-dot star of Figure 2.

We are not done with 37s. The sum of the values of the last two words of the first verse of Genesis is

$$407 + 296 = 703$$

and 703 is, amazingly, the 37th triangular number. Also,

$$703 = 37 \times 19.$$

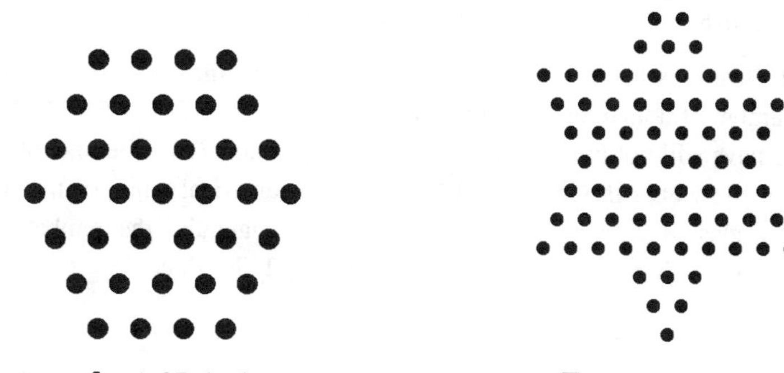

FIGURE 4 A 37-dot hexagon.

FIGURE 5 A 73-dot star of David.

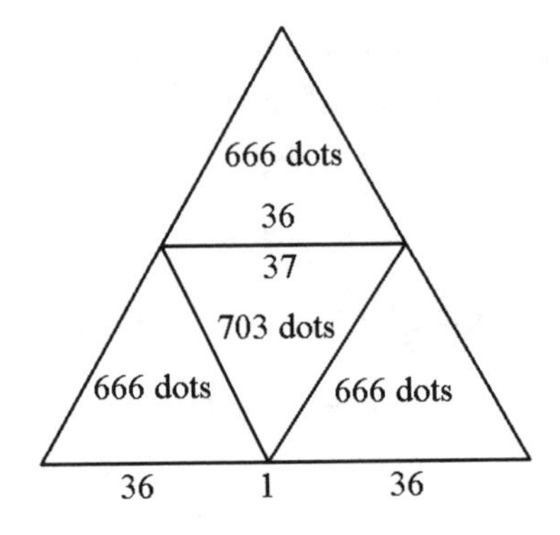

FIGURE 6 The 73rd triangular number.

When you take the 37-dot star of Figure 2 and remove six three-dot triangles you get the 19-dot hexagon of Figure 3. If you add a hexagonal perimeter to that hexagon, you have the 37-dot hexagon of Figure 4.

We are still not done. The 37th triangular number has 703 dots in it. The 73rd triangular number has 2701 dots in it. The 703-dot triangle fits precisely into the 2701-dot triangle leaving around the outside three 666-dot triangles, as shown in Figure 6. Also, 666, 703, and 2701 are the first three triangular multiples of 37.

The 37s and 73s have come thick and fast, and now here is 666, the number of the beast, and a triangular number. Mr. J. says that

> 666 is a unique triangle because all its attributes—absolute and decimal—are triangular, thus:
>
> number of sides = 3 = 2nd triangular number
> repeated digit = 6 = 3rd triangular number
> counters forming a side = 36 = 8th triangular number
> visual feature = 66 = 11th triangular number
> outline of figure = 105 = 14th triangular number;
>
> the sum of these attributes is 216, or $6 \cdot 6 \cdot 6$ this is the unique cube found earlier to be the outline of the Gen. 1:1 triangle.

I am not clear on what the "visual feature" refers to, but the "outline of figure" is the number of dots in the perimeter of the triangle, $35 + 35 + 35 = 105$.

We have still not finished with triangles. If you add 2701, the total of the values of the words in Genesis 1:1, to 302, the numerical value of the first word of Genesis 1:2, the total is 3003. Yes, it too is triangular, the number of a triangle with 77 dots on a side: $3003 = (77 \times 78)/2$. So, since

$$302 = 74 + 75 + 76 + 77,$$

the first word of Genesis 1:2 adds four rows to the triangle of the first verse and

> therefore functions as a **plinth** or **underscore**—as though to elevate further and/or confirm the high status of Genesis 1:1.

Mr. J. found a final feature that is a bit of an anticlimax: it is that the sum of the last three words of Genesis 1:1 and the first word of Genesis 1:2 is 1801, the 25th hexagonal number.

His reasoning now follows standard lines.

> These facts speak for themselves: beyond all **reasonable** doubt they are an eloquent testimony of **skilful design** and **lofty purpose**.

He denotes the author of Genesis by **A**.

> It may be further observed that
>
> - all the symmetric attributes catalogued here derive from **running sequences** of words—a fact that should not be overlooked in any assessment of probabilities;
> - clear principles are established—particularly **A's** partiality for the **unique**, the **eye-catching**, and the **symbolic**;
> - **A's** scheme of numerical design is supported on the firmest foundation known to man, namely **mathematics**—revealing an awareness that there can be no better basis for an exercise of this kind;
> - remarkably, features that specifically belong to our own system of numeration appear to have been anticipated by **A**.
> - **A's** scheme makes no great demands of the human intellect, and is open to—and no doubt, intended for—a wide audience.

The conclusion, to Mr. J., is inescapable.

A correspondent of Mr. J. asked,

> Are the apparently rare events (as they are found in Gen. 1:1) as rare as they seem, when taken along with the very many alternative events that could equally well have captured our attention?

and

> Have you any probability data to support your claims?

His answer to the first question was

> Clearly, these deserve our serious attention. That the Aleph-Tau Phenomena are firmly based upon a unique structure of pure number is demonstrated in **the Appendix** where **NUN-RESH** is identified as the **confluence** of many interesting numerical features and associations—in particular, those which directly involve the most **attention-seeking** of all numerical objects, **37**.

His breaking into boldface and capitals are symptoms of uncertainty. The appendix to which he refers contains only more of what has gone before—for example, that 370 is equal to the sum of the cubes of its digits, as is 153, another triangular number and one of Biblical significance, and in the second pair of amicable numbers, 1184 and 1210, the first is a multiple of 37.

Mr. J. chose to miss the point of the first question which was, given *any* text, would it not be possible to observe as many striking occurrences? The answer, I think, is *yes*. In the preceding chapter I had no difficulty in finding squares in the Gettysburg Address and I have no doubt I could find many triangular numbers as well. But what I found would probably not be as impressive as what Mr. J. has found, because he has, I am convinced, devoted many, many, *many* hours to the study of his numbers. I am also convinced that, given enough time and effort, I could be almost as impressive with the Gettysburg Address.

Mr. J.'s response to the second question was as follows:

> On the second matter, attention must be drawn to the kinds of evidence presented in our courts of law; these solemn verdicts are arrived at—but rarely, if ever, on the basis of probability calculations!

That is, the answer is *no*, but it could hardly have been anything else. Even an expert in probability would be hard-pressed to produce any such data. The correspondent was probably trying again to get Mr. J. to entertain the notion that what he had found was not there by anyone's conscious design.

Mr. J. also received the comment that

> what you do not make clear are your conclusions as to whether any message has been uncovered. It is difficult to understand the purpose of such extreme ingenuity unless some important message was intended.

That was an easy one, especially since Mr. J. had noticed something:

> It is an interesting and—in view of the foregoing deductions—undoubtedly a **planned** coincidence that the sum of the 1st and 3rd, and also of 2nd, 4th and 5th, characteristic values of Gen 1:1 is equal to **999**—the telephone number that, in Britain, is dialled in cases of **emergency**!

The message is that God exists, and the end times are near.

I sent Mr. J. a copy of the pamphlet referred to in chapter 10 that contained what seemed to be hundreds of 57s hidden in names, dates, battles, and so on of the American Revolution. My intention was to show him how easy it was to find numbers if you are looking for them. I did not state this with sufficient clarity, since the pamphlet bounced right off him, making no impression at all:

> I have read *The Role of 57 in the American Revolution* with interest; however, I am concerned that you appear to see this as having some relevance to my writings.

He went on

> In my view, your present attitude is being largely conditioned by a fear of the implications of my findings—though you will, no doubt, deny this.

Yes, that's right.

CHAPTER 15
Paragrams

It is the fate of minor art forms to disappear. Patter songs, chess problems with white to play and mate in six moves, six-day bicycle races and six-part fugues, light verse in rhymed couplets—gone, all gone, though they all had their days, their devotees, and their masters. This chapter will describe a minor art form, the paragram, that sprang from gematria and flourished for such a short time (less than two hundred years) in such a small space (as far as I know, in Germany only) that not only has it been entirely forgotten, it was not even widely known when it was being practiced. Though minor, and small, it provides another example of the power of numbers.

The word "paragram" is derived from a Greek phrase meaning "jokes by the letter" and originally referred to simpleminded puns such as "Biberius Mero" for "Tiberius Nero." Similar wit can be found today. In fact, it cannot be avoided. To avoid the possibility of giving offense to anyone—even Nero may have passionate defenders, ready to sue at the slightest whiff of slander—I will provide no examples, but readers can probably think of any number.

The term entered English, but just barely. *The Oxford English Dictionary* gives only one meaning:

> A play on words, altering one letter or group of letters of a word.

It then cites only three uses. The earliest is from Hobbes (1674):

> . . . paragrams; that is, allusions of words are graceful, if they be well-placed.

Then one from Addison (1711):

> Aristotle describes two or three kinds of puns, which he calls paragrams.

and one from Melmoth (1753):

> A paragram is a species of the pun which consists in changing the initial letters of a name.

Dr. Johnson's *Dictionary* takes no notice of paragrams, modern unabridged dictionaries give only the *OED* definition, and less thorough dictionaries leave it out, skipping directly from "paragon" to "paragraph."

However, paragram meant something quite different in Germany. A history of the development of the form can be found in Tatlow [1]. It had its start with Michael Stifel (c. 1487–1567), a contemporary and friend of Martin Luther, who for some reason became engrossed with gematria. He did not use what seems to us to be the obvious way of turning letters into numbers, putting a = 1, b = 2, ..., z = 26, but used instead consecutive triangular numbers:

a	*b*	*c*	*d*	*e*	*f*	*g*	*h*
1	3	6	10	15	21	28	36
ij	*k*	*l*	*m*	*n*	*o*	*p*	*q*
45	55	66	78	91	105	120	136
r	*s*	*t*	*uvw*	*x*	*y*	*z*	
153	171	190	210	231	253	276	

The sum of all the numerical values is

$$1 + 3 + 6 + \cdots + 276 = 2300.$$

Stifel therefore set himself the task of writing a poem each of whose twenty-two lines would have a sum of 2300 in his trigonal alphabet. In Daniel 8:14, 2300 is the number of days necessary to cleanse the sanctuary, all the more reason to write a 2300-poem. It is not an easy job, so it is understandable that the linguistic content of the poem's lines is not up to the perfection of the sum of their letters. Nevertheless, the poem gets its point across. Here is a translation from the Latin, adapted from [1, p. 152]:

This is the sum of sums,
the sum of sums from the alphabet.
It comes from the Latin alphabet, from numbers,
and it is the number of Daniel.
Behold the sacred sum of the whole alphabet.
The sum heard by Daniel, in the eighth of Daniel,
and the sacred sum is revealed from heaven.

Behold the number of the triangles,
it fills out the alphabet with triangles.
And behold there is a pyramid with three corners.
Behold, this is the number, this is the alphabet.
Behold this from twenty-three letters.
And from this number the calculation,
and the calculation is from single letters,
the single numbers that are the same being joined.

> Behold the alphabet: Latin, and fixed,
> the Latin alphabet in numbers.
> These same points, two thousand three hundred,
> make up the days of Antiochus Epiphanes,
> The days of Antiochus and the points of the Latin alphabet.
> These things reveal the progress of God,
> the progress, the calculation.

The reference to the pyramid with three corners is meant to bring to mind the fact that the twenty-third pyramidal number (twenty-third because there are 23 letters in the alphabet) is... 2300! The exclamation point is meant to be ironical, since it is neither a coincidence nor surprising that the number should be 2300. Summing triangular numbers gives pyramidal numbers. Think of a pyramid of cannonballs with four layers: the number of the balls in the layers (see Figure 1) are triangular numbers, and so the fourth pyramidal number is $1 + 3 + 6 + 10 = 21$, the sum of the first four triangular numbers. Similarly, the number of balls in a pyramid with twenty-three layers would be the sum of the first twenty-three triangular numbers, 2300.

From a poem with all lines having the same sum, it is a small extension to constructing parallel passages. Here is an example by Michael Poll (1577–1631) of Breslau, composed for a wedding:

JOHANNES BERNHARDUS = 1362
MAGDALENA ELISABET = 797
IHOVA HOC CONJUGIUM = 1362
IS ET BENEDICAT = 797.

(Johann Bernard, Magdalene Elisabeth: May God bless this marriage.)

[1, p. 66]

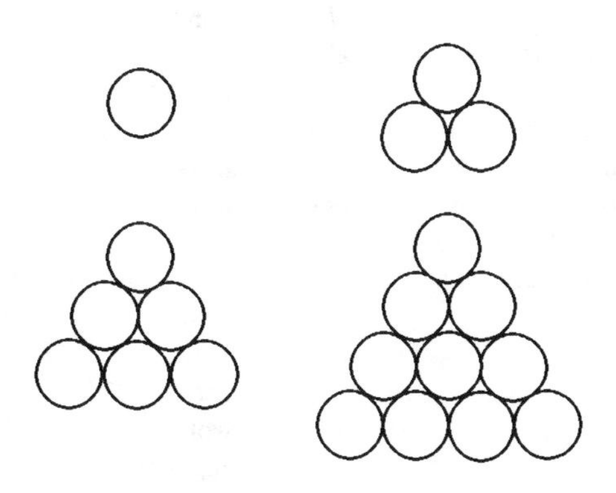

FIGURE 1 Layers of a 4-high pyramid.

In the paragram's finest flowering, an author would take an occasion, attach an appropriate Bible passage to it, and then construct a text appropriate to the occasion. The text would have to have the same gematric sum as the Bible passage. For example, here is a paragram by Johann Friedrich Riederer [1, p. 96], published in 1718, commemorating the refurbishment of a church. (Riederer's father was its pastor.) The title of the four-page publication is

Paragrammacabbalisticum Trigonale auf die neue Egidier Kirche,

Trigonal Cabalistic Paragram on the New Church of St. Egidien. "Cabalistic" refers to the translation of letters into numbers, and "trigonal" to the substitution made: the nth letter is replaced by the nth triangular number. The alphabet was slightly different from Stifel's, w having separated from u and v, so the values of the letters at the end were

uv	*w*	*x*	*y*	*z*
210	231	253	276	300.

The form of the paragram was the two passages displayed in two columns, with the numerical values for each word displayed (Table 1).

Die	70	Und	311
nunmehr	674	die	70
neu-	316	Arbeiter	575
erbaute	587	arbeiten	513
schöne	424	daß	182
Kirche	310	die	70
bey	294	Berrerung	857
Sanct	459	im	123
Egidien	249	Wercke	475
in	136	zunamhe	731
der	178	durch	415
Keyserlichen	944	ihre	249
freyen	571	Hand /	138
Reichsstadt	988	und	311
und	311	machten	417
Weltberühmten	1293	das	182
Republicq	754	Haus	418
Nüremberg	746	Gottes	699
		ganz	420
		fertig	452
		und	311
		wol	402
		zugericht	983

TABLE 1 A paragram.

At the end of each column was the total, the same for both, 9304. That is a feat and, in the days before pocket calculators, must have taken a good amount of labor to compose. One can understand why Riederer had his broadsheet printed.

The second column is 2 Chronicles, 24:13, certainly appropriate to the occasion:

> And the laborers labored so that the work of repair prospered in their hand, and they made the house of God fully ready and completely restored.

The first column, likewise appropriate:

> The now newly rebuilt beautiful church of St. Egidien in the imperial free imperial-city and world famous republic Nürnberg.

The intent of the paragram may have been only to show off its author's ingenuity but there may have been mysticism at work as well: perhaps the consonance of numbers would cause a consonance of sense or show that the restoration is in harmony with the Bible.

It's probably best not to attempt to check the arithmetic. As Tatlow points out.

> The paragrammatists often used unusual and inconsistent spelling in their paragrams as well as making occasional mistakes in their arithmetic. Printers, perhaps trying to correct what they saw as errors, sometimes introduced further mistakes. [1, p. 146]

Yes, some temptations are too hard to resist. A paragrammatist, having gotten it *almost* right, can be excused for distorting a spelling a little bit. The next-to-last word in the first column, *republicq*, may have been the way that eighteenth-century Germans habitually spelt *republic*, but I doubt it. A *little* change to make it come out right: could you resist making it when the alternative would be to make wholesale changes, in effect almost starting over? Probably neither you nor I, nor Riederer, could. Besides, everybody will know what *republicq* means.

Here is another example, by Picander (Christian Friedrich Henrici), published in 1732, written on the occasion of a wedding in 1730. There are many examples from which to choose, since paragrams were not rare. Tatlow [1, p. 12] discovered some 5800 of them, of which more than 5000 use the trigonal alphabet. The first of Picander's two passages, which sums to 20699, is

> Bey dem Daumischen und Thymischen Hochzeits Festin wolte dem Herrn Bräutigam und Jungfer Braut beyden wahre Liebe wahre Beständigkeit wahre Treu reichlich Auskommen und alles vergnügende Wohl nebst folgenden aus der heiligen Schrift ein alter Teutscher und naher Befreundter beständigst anwünschen.

That is,

> On the occasion of the Daum and Thym wedding, an Old German and close friend desired to convey his sincere wishes to the bridegroom and bride for true love, true constancy, true fidelity, abundant fortune and all pleasurable good, besides the following from Holy Scripture. [1, p. 146]

The original, being written in a column, had no more punctuation than the horizontal version above has.

The second, as is traditional, is Biblical, Sirach 26: 13–15:

> Ein freundlich Weib erfeuet ihren Mann und wenn sie vernünfftig mit ihm umgehet erfrischet sie ihm sein Hertz. Ein Weib das schweigen kan das ist eine Gabe Gottes. Ein wohlgezogen Weib ist nicht zu bezahlen. Es ist nichts liebers auf Erden denn ein züchtig Weib und nichts köstlichers ein keusches Weib.

That is,

> A kindly wife delights her husband and her good sense refreshes his heart. A wife who can be silent is a gift of God. A well-disciplined wife is beyond price. There is nothing so lovely on earth as a modest wife and nothing so precious as a chaste wife. [1, p. 146]

Paragrams seem to have died out in Germany some time after 1750 and they never caught on elsewhere. Such is the fate of minor art forms. Paragrams were an offshoot of medieval number symbolism that survived long after most other forms disappeared.

Has the time come to revive paragrams? Constructing them, or constructing anything that has to come out to a fixed total when gematrized, is not easy work, but computers could make the task considerably easier. A program to calculate the numerical value of a passage could be whipped up in short order, and any number of alphabet-to-number conversions could be included. In Table 2 is a paragram that I made, without computer assistance. I think that it reads fairly smoothly, and neither "republic" nor any other word is misspelled on purpose. It may be the first English paragram ever published. In keeping with paragrammatical tradition, it uses the trigonal alphabet with z = 300.

The occasion that it marks is the publication of this book, and the appropriate Biblical passage is from Ecclesiastes 1:1, 8. It was constructed with a little mechanical assistance from a pocket calculator, so if there are any errors in it, they are the calculator's fault.

For all its virtues and pleasures, it is probably futile to hope for the revival of the paragram. When minor art forms die, they stay dead. But the number-mystical instinct and the urge to gematrize that led to paragrams has not died. They are unkillable and, when one expression of them disappears, another takes its place, as other chapters of this book show in abundance. The spirit of Pythagoras has lost none of its strength.

Finally!	566	Emptiness,	896
Another	591	emptiness,	896
book	268	says	619
by	279	the	241
Woody	587	Speaker,	530
Dudley.	589	emptiness,	896
I	45	all	133
can	98	is	216
hardly	542	empty.	679
wait	467	All	133
to	295	things	561
read	179	are	169
it	235	wearisome;	841
all.	133	no	196
I	45	man	170
will	408	can	98
purchase	712	speak	362
many	446	of	126
copies	462	them	319
to	295	all.	133
give	298		
away.	509		
Total	8187	Total	8187

TABLE 2 An English paragram.

Reference

1. Tatlow, Ruth, *Bach and the Riddle of the Number Alphabet*, Cambridge University Press, Cambridge, 1991.

CHAPTER 16

Shakespeare's Numbers

Did you know that there is such a thing as numerological literary criticism? I never did until I picked up a copy of *Triumphal Forms*, by Alistair Fowler [1]. The book is not like numerology books, put out by obscure publishers in California, since it bears the imprint of the Cambridge University Press, a publisher of great repute and not one to publish crank literature. Its author is not like the authors of numerology books, unlearned and sometimes semiliterate, since he was a Fellow of Brasenose College, Oxford, an organization of great eminence and not noted for craziness. Nevertheless, the book contains numerology.

Dr. F. was not even an eccentric:

> An invitation to the Institute for Advanced Study, Princeton, made it possible for me to work in almost ideal circumstances of quiet and scholarly companionship. [1, p. xii]

Cranks do not get invited to the Institute for Advanced Study. Ordinary people do not get invited there, either. Only those extraordinarily eminent do, and no one attains extraordinary eminence by holding extremely odd views. But the power of numbers can overwhelm even the eminent.

Numerological literary criticism is not restricted to Dr. F.'s book. Evidently there is a whole *school* of numerological literary critics. Dr. F. wrote that

> some critics regard numerology as the key to all literary knowledge; others dismiss it as an extraneous cryptic curiosity. [1, p. 200]

The key to *all* literary knowledge! Now *there* is eccentricity. Dr. F. judiciously decided, in proper academic style, that the truth lay somewhere in between.

> Consideration will show that neither extreme position is tenable. [1, p. 200]

Perhaps. However, sometimes extreme positions are not only tenable, they are correct. "The circle cannot be squared with straightedge and compass alone" say thousands of mathematicians. "The circle not only can be squared, I have squared it" say thousands of circle-squarers. The truth about squaring the circle does not lie somewhere in between. It is right there at one of the extremes.

Numerological literary critics look for number-mystical patterns in literary works as keys to their meanings. The patterns that are found must be keys, they say, because

> [the] numerical patterns... must also have been intended, since they can scarcely occur without deliberate effort. [1, p. 201]

Yes, even as we have seen that Apollonius weighed his words deliberately before sending Euxenes away (chapter 4), as the Bible-writers intended all their 7s (chapter 12), as Abraham Lincoln must have put squares in his Gettysburg Address (chapter 13).... Since human beings are incapable of producing randomness (no one can write a sequence of 1000 digits that would not fail almost all of the tests for randomness), what they say, write, and compose contains patterns. Sometimes, very few times I think, the patterns are intended, more often they are not. Though they were not intended, they can be found by ingenious people, who then say that, since the patterns are so ingenious (ingenious people can find ingenious patterns), they must have been intended. Not so. The "must" does not hold. This is one of the main messages of this book.

As an example of numerological literary criticism, here is how Dr. F. attacked the problem of Shakespeare's irregular sonnets. Shakespeare wrote 154 sonnets of which three are irregular, namely numbers 99, 126, and 145. Most of us would think something like, "That's nice. So what?" and pass by, but not Dr. F.:

> The text of the 1609 quarto comprises 154 correctly numbered sonnets, together with a poem in 47 seven-line stanzas, "The Lover's Complaint". Not all the sonnets are metrically similar.... Besides "The Lover's Complaint", any accurate description of the sequence must take account of 3 metrically irregular sonnets, that is, 3 sonnets not in pentameter quatorzains. These are: xcix (a fifteen-line stanza), cxxvi (a twelve-line stanza in couplets) and cxlv (tetrameters). The attention previously given to these sonnets typifies the biographically fallacious approach of the textual critics. Sonnet cxxvi, which fitted the substantive division into "W. H." and "dark lady" sonnet groups, they treated as a "coda" to the former; whereas they did not regard cxlv as a structural division at all, and either ignored or explained away xcix as a "rough draft". Beeching actually speculated that the first line of xcix was an "afterthought", Shakespeare "intending afterwards to reduce" the sonnet to 14 lines; in spite of the self-referring hint in the truly extrametrical l.5, "In my love's veins thou hast *too grossly* dyed". [1, pp. 183–184]

[That is, Shakespeare is telling us, in the sonnet, that the sonnet is too long by a line—too gross.]

> The first step in any structural analysis must be to examine the pattern formed by the irregular sonnets.

The first thing we have to do is throw out sonnet number 136. Shakespeare tells us to.

> Recalling the second of the features mentioned at the outset, we may be inclined to regard a passage in cxxxvi, which has never received a satisfactory interpretation, as self-referring:
>
> In things of great receipt with ease we prove,
> Among a number one is reckoned none.
> Then in the number let me pass untold,
> Though in thy store's account I one must be,
> For nothing hold me...
>
> In their primary sense these words refer to the speaker; but they seem pedestrian, overelaborate and obscure, unless they have some further point. This I find in a secondary reference to the sonnet itself. It has to be excluded from, yet at the same time included in, the reckoning.

"One is reckoned none," so out it goes. After it is gone, there are 153 left, and 153 is the seventeenth triangular number.

> Now if we count cxxxvi out the sonnet total becomes 153, one of the best known of all symbolic numbers.... We can now discern the structural pattern governing the arrangement of Shakespeare's irregular stanzas. They are so located in the sequence that each is denoted by a triangular number within the greater triangle 153. Thus the octosyllabic cxlv begins a culminating ten-sonnet triangle cxlv-cliv, the twelve-line cxxvi begins a twenty-eight-sonnet triangle and the fifteen-line cxix begins a fifty-five sonnet triangle. Moreover, two of these triangular numbers had great arithmological significance, 10 as the principle of divine creativity itself, 28 as a symbol of moral perfection.

Dr. F. is right about the significance of 153: it is the seventeenth triangular number, it is the number of the catch of fish in Chapter 21 of the Gospel according to John, and it has other remarkable properties as well:

$$153 = 1 + 125 + 27 = 1^3 + 5^3 + 3^3$$

and

$$153 = 1 + 2 + 6 + 24 + 120 = 1! + 2! + 3! + 4! + 5!$$

I think that both Shakespeare and Dr. F. were unaware of them.

Dr. F. says that Shakespeare was thinking about triangles:

Shakespeare mentions the triangular shape in Sonnet cxxii:

> No! Time, thou shalt not boast that I do change,
> Thy pyramids built up with newer might
> To me are nothing novel, nothing strange,
> They are but dressings of a former sight...

In its primary sense, this passage refers to the obelisks excavated in Rome and re-erected (1585–90) by Pope Sixtus V; but it also glances at the form of the sequence itself, for *pyramid* and *triangle* were often synonyms. [1, p. 186–187]

Be that as it may, one might think that if Shakespeare wanted to refer to triangles he, being clever, could have found a way to do it. This raises the question of how triangular Shakespeare actually was. Assuming, as Dr. F. does, that Shakespeare was on intimate terms with triangular numbers, one might also think that a few might seep into his other work. But a search of his complete works (easily done with modern computer technology) shows that not once, ever, does he use the word *triangle*, *triangular*, or in fact any word starting with *triang*. This is so singular that it almost makes one think that Shakespeare was avoiding triangles on purpose.

His contemporary, Edmund Spenser, had no trouble with them: here is stanza 22 of canto 9 of book 2 of the *Faerie Queen*:

> The frame thereof seems partly circular,
> And part triangular; O work divine!
> Those two the first and last proportions are;
> The one imperfect, mortal, feminine,
> Th'other immortal, perfect, masculine;
> And twixt them both a quadrate was the base
> Proportioned equally by seven and nine;
> Nine was the circle set in heaven's place;
> All which compacted made a goodly Diapase.

There we have both number mysticism and shape mysticism, all in nine lines.

Shakespeare seems to me to have been one of those persons insensitive to numbers and their power. In all of his works I could find only six references to "number" where the word was used in anything other than the prosaic sense of something to count with. One, from *The Phoenix and the Turtle*, though touching, is a fairly obvious image and has nothing of mysticism about it:

> So they loved, as love in twain
> Had the essence but in one;
> Two distincts, division none;
> Number there in love was slain.

Another (*Love's Labour's Lost*, Act 4, Scene 3) is not deep:

> Dumain. Now the number is even.
> Berowne. True, true, we are four.

The third (*The Merry Wives of Windsor*, Act 4, Scene 1) is, I think, a joke:

> Evans. William, how many numbers is in nouns?
> William. Two.
> Quickly. Truly, I thought there had been one number more, because they say "Od's nouns."

Sonnet 38 contains a reference that is not mystical:

> And he that calls on thee, let him bring forth
> Eternal numbers to outlive long date.

Sonnet 79 has

> But now my gracious numbers are decayed.

but "numbers" here is being used as a synonym for "poetry," or "verse," perhaps because neither of those words would fit the meter.

The one and only trace of number mysticism in all of Shakespeare is in *The Merry Wives of Windsor*, Act 5, Scene 1, where Falstaff says

> ... they say there is divinity in odd numbers, either in nativity, chance, or death.

This echo of primitive neopythagoreanism is not what would be called deep knowledge of the inwardness of integers.

The conclusion to be drawn is that Shakespeare did not care about number mysticism, or about numbers. It is impossible to prove a negative proposition like that, but I think it is just as difficult to prove that he knew anything at all about the properties of integers beyond their use in counting.

This was not Dr. F.'s conclusion. Since 153 is a triangular number, it is natural to look, he said (implying that Shakespeare must have also done so), at the triangular arrangement:

154
152 153
149 150 151
145 146 147 148
140 141 142 143 144
133 134 135 137 138 139
126 127 128 129 130 131 132
118 119 120 121 122 123 124 125
109 110 111 112 113 114 115 116 117
99 100 101 102 103 104 105 106 107 108

There they are! The irregular sonnets, all starting lines of the triangle (with 136 missing), spaced uniformly. What is even more significant about sonnet 126 is

> Self-referring ambiguity again occurs in cxxv, a stanza difficult to interpret satisfactorily, unless it applies in part to the sequence's external form:
>
> > Were it aught to me I bore the canopy,
> > With my extern thee outward honouring,
> > Or laid great basis for eternity,
> > Which proves more short than waste or running?
>
> The very next stanza is the "short" cxxvi, beginning the base of the pyramid of 28. We conclude that the "basis for eternity" described as more short than waste is in one sense Sonnet cxxvi itself, the sonnet on Love defying Time. Time, which figures very prominently in earlier sonnets, is never mentioned in the sequence after cxxvi. Moreover, from a numerological point of view, 28 is suitable to eternity, as a number symbolising the perfect bliss in heaven towards which saints yearn. [1, p. 187]

That is, Shakespeare decided—he thought it out beforehand—to stop referring to time twenty-eight sonnets before the end of the sequence because twenty-eight is a triangular number, and also perfect. Is this credible? Yes, says Dr. F., it is:

> The pyramidal numbers imply, most obviously, that Shakespeare designed the sequence as a monument. [1, p. 188]

If Shakespeare was designing a monument, you might think that he would have taken his design all the way to its bottom. But Dr. F. says nothing about sonnet number 63, which would be the number in the triangle three rows below number 99. For consistency, it too should show signs of irregularity, as should number 18, three more lines down.

The number of sonnets is very close to $156 = 3 \times 52$. Dr. F. says that this is no accident. Shakespeare meant there to be three 52-week sections.

> At first, Sonnet lii, with its reference to "the long year", seems to fall into the same trivial category. But then we notice that a further year of 52 stanza-weeks would bring us to civ, a sonnet entirely devoted to meditation on the "process of the seasons" and to reckoning years passed away since the poet first saw his beloved. It seems as if the sequence as a whole may be ordered according to a modulus of 52 stanzas per year, so that a division into 3 year-parts (i–lii, liii–civ, cv–cliv) constitutes yet another threefold arrangement. [1, p. 192]

The question is, who is the master here, the sonnets or the numbers? I think it is the numbers. I think that Dr. F. first decided that the sonnets ought to be divisible into three 52-week parts and then started to look for evidence to support that conclusion. The alternative is that he was, out of the blue, struck with the strong signals that the text was sending out that it was meant to be taken in three parts. The signals do not seem to be that strong.

Also, inappropriate signals are ignored. For example, in sonnet 73 we have

> That time of year thou mayst in me behold,
> When yellow leaves, or none, or few do hang
> Upon these boughs which shake against the cold, . . .

Brrr! In the sonnet it is November, or December. But $73 - 52 = 21$, and the twenty-first week of the year comes towards the end of May. If the sonnet subsequences were mimicking the year, this one was misplaced and should have had a number nearer to the end of the year. Dr. F. does not mention this sonnet. Nor does he mention the line in sonnet 53 (happy New Year!),

> Speak of the spring, the foison of the year.

Selecting evidence to support a theory is a scientific sin. But scholars of English who have not had a solid grounding in the scientific method cannot be held to the same high standard as, for instance, chemists. Especially when numbers are involved.

There remains the difficulty that 154 is not 156. This must be explained.

> . . . this still leaves us 2 short of the expected 52. In other words, for our hypothesis to hold there will have to be 2 missing sonnets or, at least, 2 places where the sonnet total is increased without addition of actual sonnets. Now Sonnet lxxvii with its mentions of "blanks" seems to hint at a similar possibility: "Look what thy memory cannot contain,/ Commit to these waste blanks . . . " Editors usually give a biographical explanation of these lines, perhaps speculating that Shakespeare wrote lxxvii as an occasional sonnet to accompany the gift of a book with blank pages. Some such interpretation is no doubt right. But may we not also take "blanks" in a secondary sense as referring to blanks within the sequence itself? May not the blanks, in short, be blank sonnets, numerological makeweights like the blank chapters in *Tristram Shandy*? [1, p. 192–193]

There *have* to be blanks. The numbers demand it. What numbers demand, numbers will get. When it comes to a choice between numbers and reality, the numbers win.

Dr. F. found the blank sonnets. His book includes a picture of the 1609 folio showing sonnet 126, the short one with only twelve lines, with two sets of parentheses following it:

> . . .
> Her Audite (though delay'd) answer'd must be,
> And her Quietus is to render thee.
> (   )
> (   )

It seems to me that there is a simple explanation for the parentheses. The printer had finished setting fourteen-line sonnet after fourteen-line sonnet, and here was something with twelve lines. The next sonnet had fourteen lines, as did the one after that, the one after that, and so on. If you were a reasonably intelligent printer, would not the thought cross your mind that the copy you were setting had two lines missing? And that the author, reading proof, would see it, smack himself on the forehead, mutter a curse or two, and indicate the missing lines in the margin? And what would happen if you, the printer, did not allow space for the two missing lines? The *whole* of the work from there on might have to be reset. How intelligent it would be to eliminate that extremely tedious work by leaving space for the missing lines, and by putting in parentheses so that the place where they are missing can be found quickly.

That is not how Dr. F. saw it.

> Scholars have assumed that the 2 sets of parentheses between cxxvi and cxxvii indicate the 2 lines by which the former stanza falls short of the fourteen-line sonnet norm; and perhaps incidentally they do serve this unusual purpose. But, if so, they perpetrate a numerological and typographical pun. For, as a little reflection will show, their primary function is much more likely to be the important one of marking the "waste blanks" referred to in lxxvii and more recently in cxxv. [1, p. 193]

No, a little reflection will not show that. A little reflection will make one wonder how a missing line can stand for a missing sonnet, and why Shakespeare didn't write 156 sonnets in the first place. He certainly would have had no difficulty in turning out two more. That would have obscured the triangularity of 153, if he was aware of it, but that could easily have been gotten around by lettering instead of numbering the last two sonnets, or calling the last three $153'$, $153''$, and $153'''$, or something similar. That would make the inner structure of the sonnets clear.

But students of literary numerology do not find such obvious marks. The ones they find are, instead, subtle in the extreme. There are two explanations for that. One is that the authors were indeed subtle in the extreme, and thought it worthwhile to make the tremendous effort to leave clues whose chance of being found and properly interpreted was minuscule. The other explanation is that the literary numerologists, people with good minds but not enough matter to work on, are seeing things that are not there.

> The question may be asked, Why only 2 blank sonnets? Would it not have been more economical of effort to combine the 153 and $3 \times 52 = 156$ totals by having 3 blanks? Perhaps; but we should then have missed the beautiful effect whereby the poet reduces himself formally as well as verbally to nothing, in cxxxvi—"For nothing hold me". [1, p. 193]

Dr. F. closes with more symmetries. If we add in the 47 stanzas of *The Lover's Complaint*, bound in the 1609 folio with the sonnets, we can divide the corpus as follows:

> **52** sonnets
> **24** more sonnets
> 26 more sonnets, with **365** lines
> **24** more sonnets
> the last 28 sonnets and the first 24 stanzas of "TLC", **52** stanzas
> the final 23 stanzas of "TLC".
>
> The theme of time hours, days, years is apparent, as is the symmetry of $a\ b\ C\ b\ a\ d$. [1, p. 196]

Dr. F. does not comment on that nagging final 23.

Shakespeare may have composed his sonnets with all this in mind, but I have the feeling that Dr. F. is getting more out of Shakespeare's works than Shakespeare put into them. Many artists have put number patterns into their work (Dürer's famous *Melancholia*, with its magic square giving the date of its composition; J. S. Bach working an amazing amount of arithmetic into his music), but I think it was for the fun and challenge of doing it and not as secret keys to meaning. When critics use number mysticism to get at reality, either physical or literary, they slip into numerology and go too far. The reason they slip, I think, is that they have active minds, and good ones, with not enough matter to engage them. They thus have to make some up. Mathematicians are much luckier than literary critics: there is plenty of matter, at all levels, to wrestle with, and it will never give out.

Reference

1. Fowler, Alistair, *Triumphal Forms*, Cambridge University Press, Cambridge, 1970.

CHAPTER 17
Rithmomachy

Anyone for rithmomachy? Probably not, but then no one is against it either, because rithmomachy has been entirely forgotten. Rithmomachy is a game, the Battle of Numbers, unlike any game played today but one that was played for more than six hundred years. Its origins are not clear, but it is clear that it is one more example of what Pythagoras wrought, and one more example of the power of numbers to fascinate. It is also an example of how different medieval minds were from modern minds.

In its day, rithmomachy was at least as important as chess, though chess has survived and rithmomachy has not. Pythagoras had nothing to do with chess, which originated in the East, evolved into more or less its present state in India around 600 A.D., and then traveled west. Chess has no numbers in it but, as we will see, rithmomachy is almost all number. One reason rithmomachy died out may be that its players had to be quite good at arithmetic, whereas no schooling at all is necessary to excel at chess. Another may be that it is, to modern eyes, a very strange game. Or it may have been just that fashion changed, which seems to happen for no reason at all. Games come and go: canasta, once very popular, has disappeared, and even Trivial Pursuit may some day pass into oblivion.

Rithmomachy had a good run. It was first written about in the eleventh century. Treatises were written about it in Latin, French, Italian and German; it appears in manuscripts from the eleventh, twelfth, thirteenth, and fourteenth centuries and in printed works from 1482, 1496, 1554, 1562, 1572, 1616, and 1705, though the game seems to have more or less died out by 1600. More information on rithmomachy can be found in references [1–4].

During its six hundred years of life, many rithmomachy sets must have been made and sold, and people must have spent many hours with the game, some pleasant and others no doubt passionate. Perhaps there were rithmomachy addicts, who would spend all of their time on the game if they could, and perhaps there were rithmomachy professionals who made a living by traveling around, challenging and beating local rithmomachy experts. Or perhaps not: the medieval mind was different from ours, and we should not try too hard to project ourselves into it. We do not know exactly the place of rithmomachy in medieval times because historians often do not mention the truly important features of daily life, like games, concentrating instead on tedious political maneuvering.

Of course, the historians are not entirely to blame because the chroniclers of a time often do not mention these things either. There are several questions about the game whose answers it would be nice to know. First, how in the world could the game ever have been invented? Second, how did it become popular enough to survive for as long as it did? Third, is it fun to play? We will probably never know the answers to the first two questions, but we can investigate the third. I will describe the game, and I encourage you to try it out. After you do, if nothing else you can correctly claim to be the best rithmomachy player in your town. Anyone for rithmomachy? Perhaps *you* are.

Rithmomachy is played, by two people, on an 8-by-16 checkered board. Thus, the fact that the last maker of rithmomachy sets went out of business more than three hundred years ago need not stop you from playing the game: two chess boards placed end-to-end are all that you need. The pieces, though, will take some work. Each player has twenty-four pieces. Black has the odds, and White has the evens. The odd side has pieces numbered

3	5	7	9	9	12	16	25
28	30	36	49	49	56	64	66
81	90	100	120	121	190	225	361

while the even side has pieces numbered

2	4	4	6	6	8	9	15
16	20	25	25	36	42	45	49
64	72	81	81	91	153	169	289

Each side has eight circles, eight triangles, seven squares, and one pyramid. The numbers on the pieces may seem arbitrary and irrational (though they are integers), especially since even numbers occur on the odd side and odd numbers on the even side, but they are not. Though it does not leap to the eye, there is logic and symmetry behind them. The odd pieces all come from 3, 5, 7, 9 in the same way that the even pieces come from 2, 4, 6, 8, as Table 1 shows.

As the formulas show, the numbers are not arbitrary; they are almost inevitable. Rows 1, 2 and 4 are the fundamental rows. Once you have the idea of starting with a number, its square, and the next square in rows 1, 2, and 4,

	Odds				
Circles	3	5	7	9	x
Circles	9	25	49	81	x^2
Triangles	12	30	56	90	$x(x+1)$
Triangles	16	36	64	100	$(x+1)^2$
Squares	28	66	120	*190	$(x+1)(2x+1)$
Squares	49	121	225	361	$(2x+1)^2$
	Evens				
Circles	2	4	6	8	y
Circles	4	16	36	64	y^2
Triangles	6	20	42	72	$y(y+1)$
Triangles	9	25	49	81	$(y+1)^2$
Squares	15	45	*91	153	$(y+1)(2y+1)$
Squares	25	81	169	289	$(2y+1)^2$

TABLE 1 Rithmomachy pieces.

the other numbers follow by addition. Row 3 is the sum of rows 1 and 2, row 5 is the sum of rows 3 and 4, and row 6 is the sum of rows 2, 3, and 5.

The pieces marked with the asterisks are the pyramids, each made up of other pieces. The pyramid numbers, 190 and 91, are special because

$$190 = 8^2 + 7^2 + 6^2 + 5^2 + 4^2$$

and

$$91 = 6^2 + 5^2 + 4^2 + 3^2 + 2^2 + 1^2.$$

The 190 pyramid is made up of two squares, 64 and 49, two triangles, 36 and 25, and one circle, 16. The 91 pyramid is two squares, 36 and 25, two triangles, 16 and 9, and two circles, 4 and 1. We lack complete symmetry since 190 and 91 appear in different places in the odd and even charts. This cannot be helped since the number appearing in the place corresponding to 190, 153, is unfortunately not a sum of consecutive squares, nor is 120, the number corresponding to 91. It would be nice if both pyramids had the same number of pieces in them but this is also impossible. We can't have everything.

The pieces are arranged on the board as shown in Figure 1. The players, by the way, are supposed to sit facing the long side of the board, to make the pieces easier to reach.

After we know how the pieces move, what they do, and what the object of the game is, we will be ready to play. Circles move one space in any direction, triangles move two, and squares move three, as long as the intervening spaces are empty. Pyramids move as squares.

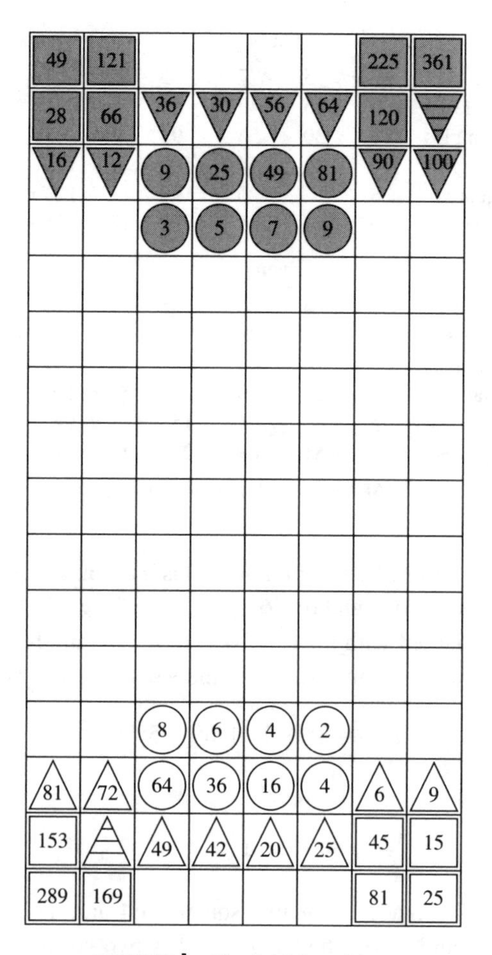

FIGURE 1 The initial position.

The shapes of the pieces were meant as mnemonics: in the old way of counting, the triangles moved three spaces, where the square they were on was counted as number one, and the squares moved four, so a glance at their shapes recalled how far they could go. Figure 2 shows examples.

Since rithmomachy is partly a war game, pieces can capture each other. Here are four methods, translated from a 1556 work on the game by Claude de Boissière [2, p. 200]. The first is the encounter:

> An encounter: if one number finds another of the same value occupying the space to which it is able to move, it will take it from the board and will move to its position.

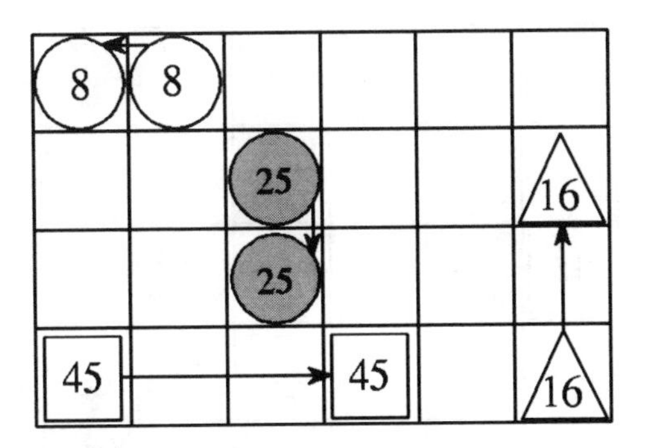

FIGURE 2 Movements of pieces.

The only Black and White pieces with the same numbers are 9, 16, 25, 36, 64, and 81 so not many pieces can be captured this way. Figure 3 illustrates captures by encounter.

The captures were modeled on medieval war. Besides random encounters of enemies, defenders of a besieged fortification could sally forth to attack the besiegers:

> A sally: a smaller number will be allowed to attack and intercept one that is larger, if it becomes equal to it when multiplied by the spaces lying between it and the larger number. And it can do this by moving in a straight line or at an angle, for every movement of at an angle, for every kind of movement is appropriate to this form of capture.

Only the empty spaces count. The restriction of the movement of the pieces to one, two, or three squares does not hold for sallies. Thus, it would be possible for the circle 9 to capture the square 81 if there were nine empty spaces between them. Similarly, the circle 16 could capture the triangle 64 by moving five squares instead of its customary one. However, the triangle 30

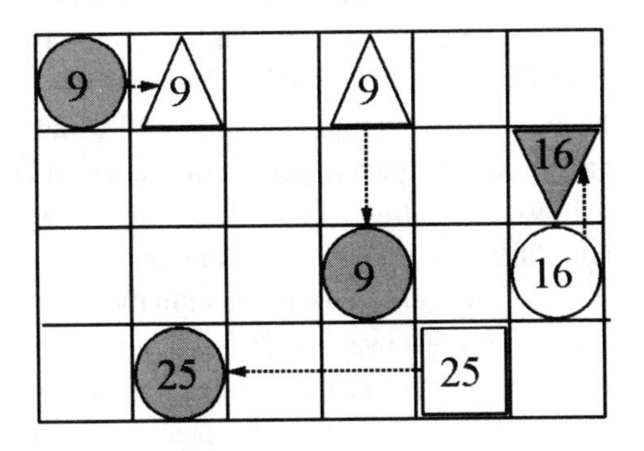

FIGURE 3 Captures by encounter.

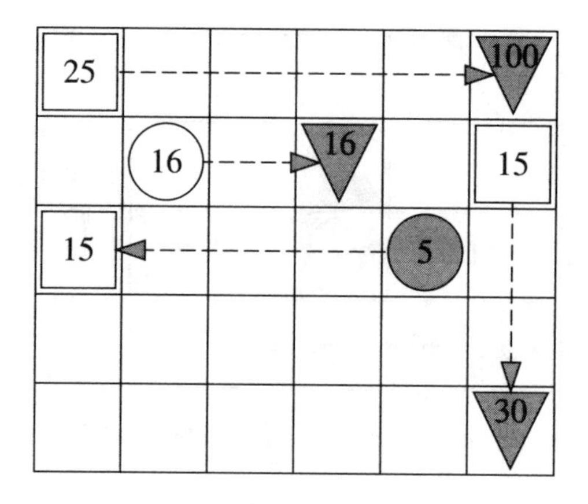

FIGURE 4 Capture by sally.

is forever safe from the circle 2 since the board is not long enough for there to be fifteen empty spaces between them. (If rithmomachy ever regains its former popularity someone will no doubt introduce hyperrithmomachy, with *three* adjoining chessboards; then 9 could take 153 and 6 could capture 120.) There is an advantage in having the even numbers when making sallies since the evens have 21 possible captures and the odds have only 17.

Next comes another technique of medieval, and modern, war.

> An ambush: two smaller numbers can ambush a larger one, if they are so placed that their united strength can stop the attack of a larger enemy of the same value; thus it happens that a larger opponent succumbs to the trickery of the small numbers.

That is, if two pieces can both be played by legal moves to a square occupied by an enemy piece and if their sum is equal to the number of the enemy, then the enemy piece is captured and one of the two attackers (either one) takes its place. The rules of rithmomachy were not as fixed as are the rules of Monopoly or chess, and some writers allowed more than two pieces to attack while others allowed multiplication as well as addition. But these were clearly later additions designed to pep up the game, and it is best to start with rithmomachy in its original pure state, or as close as we can come to it. Some ambushes are illustrated in Figure 5. In the first, the circle 5 moves and combines with the circle 3 to ambush the circle 8. In the second, 49 moves and 169 is ambushed. In the third, 9 moves and 25 is captured. After the ambushing move, either of the pieces can replace the captured opponent.

Again it is an advantage to have the even pieces, since there are thirteen possible ambushes for the evens, from 4 and 8 capturing 12 to 72 and 289

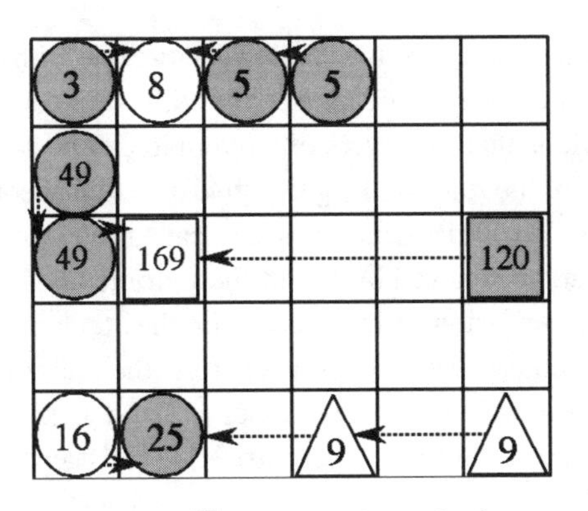

FIGURE 5 Captures by ambush.

combining to destroy 361, but only twelve for the odds: $3+5 = 8, 3+12 = 15, \ldots, 64 + 225 = 289$. To play rithmomachy you need to have a grasp of addition and multiplication, a set of capture tables to refer to, or a good deal of rithmomachial experience. Rithmomachy is not a game for children.

Medieval war also had sieges.

> A siege: when numbers are besieged by others in such a way that they cannot save themselves either by flight or by the help of their companions, they are forced to surrender to the enemy when they have been cut off.

That is, if a piece cannot move because of the enemy, it dies. This is the only method that can be used to capture pieces numbered 2, 3, 4, 5, 6, 7, 153, and 190. Figure 6 shows the unfortunate triangle 56 falling to a siege. It would likewise fall if the besieging pieces were each withdrawn one square from the triangle, since it would still be unable to move.

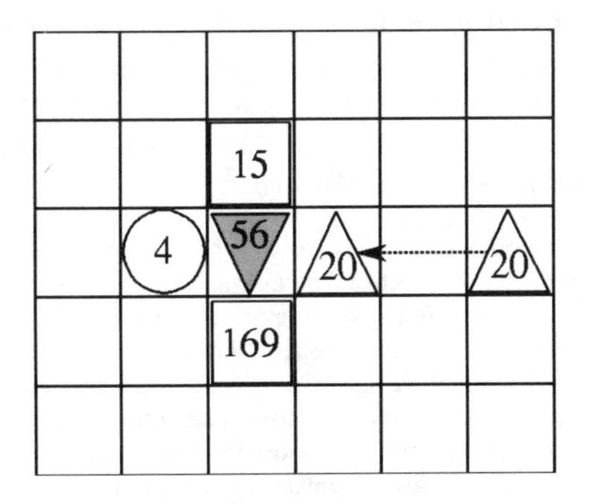

FIGURE 6 A capture by siege.

Pyramids are special cases. It is hard to capture one with the four methods: 190 could be taken only by siege, and 91 only by siege or by ambush by 25 and 66. So, the rule is that the layers of a pyramid can be separately attacked. Different sources for the rules of the game give different accounts for what can be done to a pyramid, but it seems most logical to say that any layer (or the whole pyramid) can be attacked by any of the four methods.

However, the confusion of the sources implies that any rule that players want to agree on is all right. As mentioned before, the rules of medieval games, like the rules of medieval spelling, were not as fixed as are present-day rules and orthography, and variation was permitted. De Boissière says:

> You must make a board at least ten squares long and eight squares wide; but if its length is extended to sixteen squares, the resulting increase in numbers will be an improvement.

As the size of the board was not fixed, neither was the movement of the pieces:

> First of all, the pieces will move in a straight line, either to the right or the left, forward or back; but they will not move at an angle, like the mad warriors in chess, unless the players decide otherwise and permit something else.
> [2, p. 178]

If fundamental matters like the size of the board and the movements of the pieces were fluid, it is no surprise that smaller items were subject to variation. It will take experimentation to find the best pyramid rules. There is great scope here, and after *Rithmomachy World* starts publication it will not lack for contributions, nor will the more scholarly *Journal of Rithmomachy*.

After seeing how the pieces move around and capture, the question is, how is the game won? Rithmomachy is unusual in having more than one way to win. Rithmomachy is more like a whole set of games, with different ones suitable for different levels of players. There were the five *Common Victories* and the three *Proper Victories*. Presumably beginners would start with the common victories and work up. As de Boissière said,

> We shall now describe the various kinds of victories, for we wish to arouse the enthusiasm of the players, so that they may achieve the honor of a victor's crown and the winner may not be disappointed in his hopes. Though the common victory does not really belong to this game as we have described it, yet as we want to help beginners to become more experienced and to understand the special victories, we must explain this kind first, since it is easy. Meanwhile I shall ask the experts to excuse me for descending so far as to explain the common victory, as I am interested in the progress of beginners and those who are ignorant of the game. [2, p. 204]

This implies that there was a rithmomachy community of significant size in the 16th century. Rithmomachy was perhaps the chess of its day, and it may be no coincidence that its decline and the ascent of chess occurred at about the same time. It may be that societies can support no more than one purely strategic board game at a time. Today the West has chess and the East has go, and Western go players are in general as weak, or as strong, as Eastern chess players.

In any event, in rithmomachy the simplest victory was earned by the first player to capture a number of pieces agreed on before the start of the game. Four was the number given as an example by de Boissière, and it is probably a good number to start with to become familiar with the moves and patterns of play.

The next victory—the next game, really, since its object is different—is won by the first player to capture pieces whose total is a number specified in advance. It is not clear whether the number had to be arrived at exactly, or whether it could be exceeded. The second choice would seem to make for the easier game, since to play to capture pieces summing to 124 exactly would require tables of possible winning captures, or pauses for calculation. It would also distort the game by making some pieces uncapturable, for example the 30 triangle if you were aiming for 124 and had already captured the 100 triangle.

The third victory is won in a game where the object is to equal or exceed a given total by capturing a fixed number of pieces. For example, if the game were 100 with 8 pieces, a winner could capture pieces numbered 3, 5, 7, 9, 9, 12, 25, and 30 or, on the other side, pieces numbered 2, 4, 4, 6, 8, 15, 16, and 45. It does not matter if the total is greater than 100, but it is not a win if seven or fewer of the pieces have a total greater than 100. All eight must be necessary.

In the last of the common victories, the total, the number of pieces, *and* the number of digits on the captured pieces are all specified. Here we are getting close to specifying a specific list of pieces to be captured, since there will not be many different ways to win a game with all three of those restrictions. It is only the number of digits that must be hit on the nose: you may have a higher score, or take more pieces, than the numbers specified. For example, to win the game of 160 total, 5 pieces, and 9 digits, you could take 5, 25, 30, 36, and 64 from the odd pieces, or 2, 16, 36, 42, and 64 from the even. It would be an interesting computer exercise to tabulate all possible games and how they could be won.

Good games! Rithmomachy is like chess in being a game of pure logic, and it is better than chess in that it is many games, so there would never evolve the standard opening play that many chess players spend hours (or, some of them, months) memorizing. If the players change the sum to be captured from

160 to 154, or change the number of pieces to be captured from 5 to 6, then you have a whole new game that the better rithmomacher, rather than the better memorizer, would win.

After beginners have mastered the common victories, they are ready for the proper victories.

> Now that the common victory has been fully explained, the proper victory must be described. This means the arrangement in the enemy's lines of numbers that are unlike each other, but are united in a certain proportion. And this is called a proper victory with good reason. For just as those who make war win a victory after routing the enemy's battle-line and arranging everything in accordance with their own will, so too the harmony and arrangement of numbers, which make such a noble and remarkable display of proportions inside the lines of the enemy, present the player who controls them with a most glorious victory. And this victory will be found in three forms; the great, the greater, and the excellent. [2, pp. 206–207]

The great victory consists in placing three pieces in a line, presumably next to each other, though de Boissière does not say that, the numbers on the pieces being in arithmetical, geometrical, or harmonic progression. Even could thus win with 2, 4, 6 (arithmetical), 2, 4, 8 (geometrical), or 9, 15, 45 (harmonic—$(1/9+1/45)/2 = 1/15$). The greater victory occurs when there are four pieces arranged so that two sets of three make two different kinds of progression. For example, 9, 81, 153, and 289 contains the arithmetical progression 9, 81, 153 and the geometrical progression 81, 153, 289. The excellent victory is gained when four numbers are placed that contain three subsets of three numbers exhibiting each of the three progressions. De Boissière says

> Since these are united in such a remarkable way, they cannot fail to provide an amazing and delightful spectacle, which the players and spectators can observe and enjoy with the greatest pleasure. [2, p. 211]

De Boissière gives eight sets of four numbers that give the excellent victory:

2	3	4	6
3	5	15	25
4	6	8	12
4	6	9	12
5	9	45	81
5	25	45	225
6	8	9	12
12	15	16	20

Only (4, 6, 9, 12) contains what we would call a geometric progression—4, 6, 9. None of the others are like that. De Boissière broadened the definition of geometric progression to include any four numbers which have two pairs with the same ratio, as $4/2 = 6/3$, $16/2 = 72/9$, and $15/3 = 25/5$ in the first three quadruples. Also, in all of the quadruples pieces from both sides occur, so to achieve the excellent victory it would be necessary to somehow force some of the opposition's pieces to move to where you wanted them and then immobilize them. Excellent victories were probably rare.

To play rithmomachy, skill at arithmetic is necessary, so during the middle ages its play was restricted to the small class that had that skill: those at schools and those who had passed through them. Thus, rithmomachy was primarily a game of the clergy. The nobility, innocent of geometrical progressions, played chess instead, since to checkmate a king no arithmetic is necessary. Today chess enjoys high esteem as an intellectual game, so it is amusing to think that a few hundred years ago there were no doubt many old codgers deploring the spread of chess, that crude and brutal game (with "mad warriors"—the knights—making their erratic moves), at the expense of the more refined and cerebral rithmomachy. Thus does the world, in every generation, manage to provide fresh evidence of its decline and decay.

The origins of rithmomachy are unknown. Artmann says

> The inventor of the game seems to have been a cleric around 1030 named Asilo of Würzburg. The next person to write about the game is Hermann the Lame around 1040 from the Abbey of Reichenau at Lake Constanz. He in turn is followed by an anonymous writer from Lüttich (Liège in Belgium), which was "the Athens of the North." (The mathematicians from Lüttich were the first ones to actually calculate with arabic numerals.) Fortolf of Bamberg in 1130 gave a comprehensive version of the rules and even composed some tunes based on the musical harmonies (= proportions) of the Rithmomachia. The basic motive of Asilo seems to have been the intention to familiarize his contemporaries with the *Arithmetica* of Boethius by way of the game, and this had been accomplished surprisingly well by the time of Fortolf.
>
> [1, pp. 77–78]

But it does not seem likely that so elaborate a game sprang, all new, from the head of one person. Bridge came from whist, chess was once played with dice, and even Monopoly had a more primitive predecessor. Rithmomachy is one more thing that Pythagoras wrought. Its name is Greek: clearly, with its arithmetic and geometric progressions, it derives from Pythagorean number theory. The numbers on the pieces are mostly Pythagorean square and pronic numbers. Rithmomachy is a Pythagorean game.

It may be that Asilo invented the game, all by himself, but I think it is more likely that he was elaborating some already existing game, of which we have no record. An ancestor of rithmomachy may have been played by Alexandrian Neopythagoreans or Byzantine Neoneopythagoreans and then spread to Europe.

Another question is, is it fun to play? The only way to find out is to try. Perhaps it can be revived, making chess's preeminence only a temporary aberration. But, alas, rithmomachy is more likely to stay dead, since its time, and the time of the people who would play it, has passed. It was a strange game, and its players must have been strange people. Not so strange, though, that they were not drawn to numbers. Numbers are always attractive. People change, but numbers are constant.

References

1. Artmann, Benno, Book review, *Mathematical Intelligencer* **11** (1989) #3, 77–79.
2. Richards, John F. C., Boissière's Pythagorean game, *Scripta Mathematica*, **12** (1946), 177–217.
3. ———, A new manuscript of a rithmomachia, *Scripta Mathematics* **9** (1943) 87–99, 169–183, 256–265.
4. Smith, David Eugene and Clara C. Eaton, Rithmomachia, the great medieval number game, *American Mathematical Monthly* **18** (1911) #4, 79–80.

CHAPTER 18

Number Forms

Do you have a number form? Probably not, since only around five percent of the population has one. However, you may have one and not know it, since considerably less than five percent of the population know what a number form is.

A number form is a shape that some people have inside their heads on which, or in which, numbers *are*. If that is not clear, perhaps these examples (Figures 1–4) of number forms will help.

These are copies from sketches made by the possessors of the forms, students in my classes who responded affirmatively to the question, "Does anyone here have a number form?" They are not as elaborate as some number forms that have appeared in print, for example as in Figure 5, but, just as most people are ordinary, most number forms are ordinary.

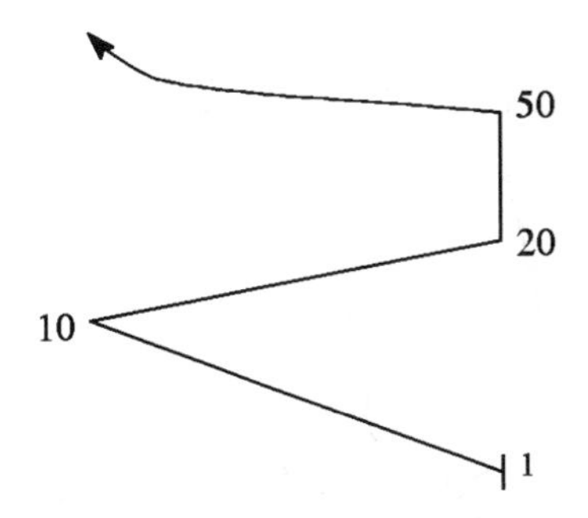

FIGURE 1 A typical number form.

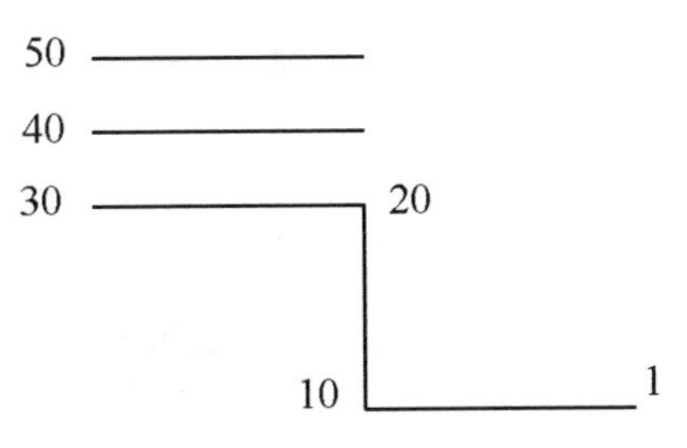

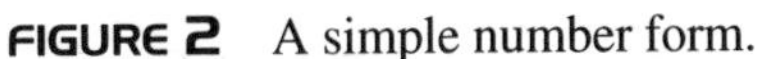

FIGURE 2 A simple number form.

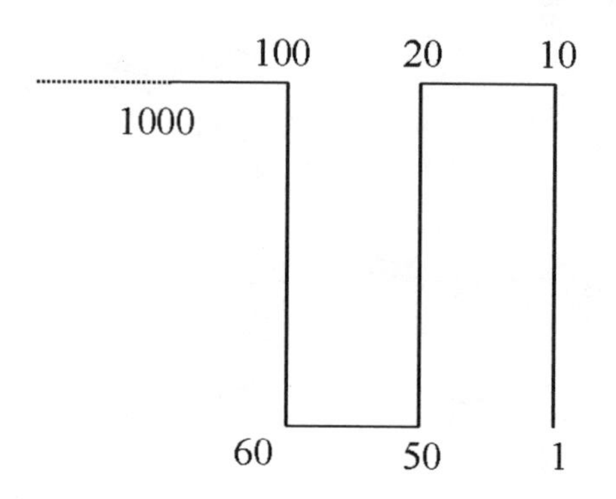

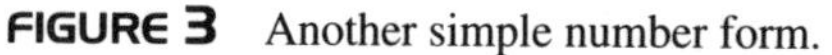
FIGURE 3 Another simple number form.

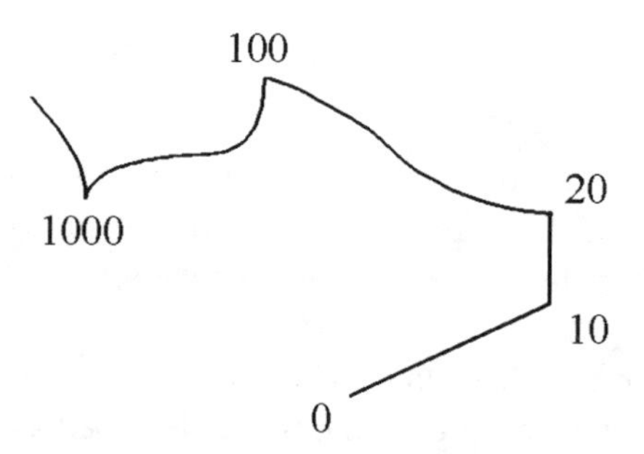

FIGURE 4 A curved number form.

Pictures of number forms are necessary, words being inadequate to describe them. One possessor of a number form tried by saying,

> The number form goes from bottom to top as far as a hundred and then from bottom to top again (and from right to left) but never coming quite as far down as before. Also, the space taken up by the first twenty numbers is greater than for any other group of twenty numbers.

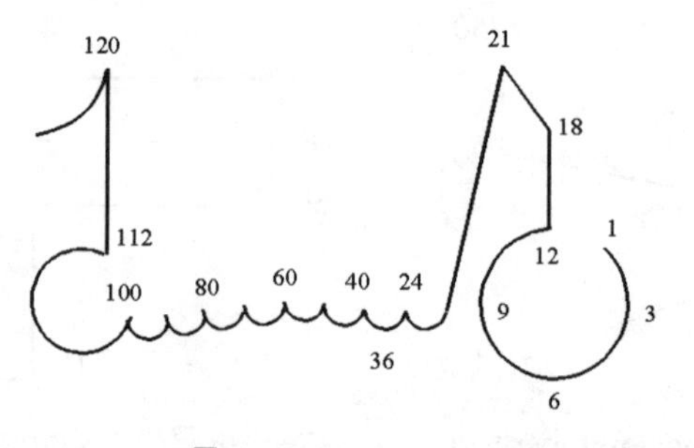

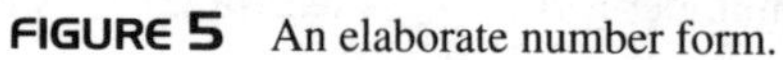
FIGURE 5 An elaborate number form.

The description is perfectly clear to the person with the number form, but not so clear to everyone else. Sometimes, even pictures are not enough:

> I also see my numbers in colors. The first ten digits have a specific color with all their multiples by ten having the same color.

Number forms do not change with time, nor do they disappear. People with number forms all say that they have had them for as long as they can remember. They are permanent parts of peoples' brains. You cannot tell if a person has a number form by looking at them, and people with forms are indistinguishable in any way from the common run of formless humanity. People who have number forms are usually no better at dealing with numbers than those who lack them. People with number forms are people who have number forms, and that is all there is to it.

Thus number forms cannot be blamed on Pythagoras, or on anyone else. They are evidence of the power of numbers over, or in, our minds. People with number forms did not decide that they wanted a number form and created one. They had no choice in the matter. Their number forms *are*.

Number forms are evidence as well that numbers do not exert power over everyone in the same way, or to the same degree. Most people do not have number forms, but there are those that are susceptible, and they do. There are some people over whom numbers seem to have no power at all, but I am convinced that when you take the race as a whole, it is true that it has an affinity for numbers. That is why what Pythagoras started continues to flourish today.

The affinity the mind has for numbers also partly explains why they make up so large a part of the education of our children. Algebra is studied in every high school, though its usefulness to almost all those who study it is as close to zero as makes no difference. The number of students put to calculus in the Unites States each year approaches three quarters of a million. 750,000! Calculus, for heaven's sake! The mind is staggered at the absurdity of it, until the mind remembers that the human brain is so constructed as to be captivated by numbers. We are made for numbers. Number forms are evidence.

If you have a number form, you know what I am talking about. In fact, you may be surprised to discover that *everyone* doesn't have one. I knew one such person: he had a number form and he could not understand how people could get along without one. He told me that when he told his wife about his number form, she looked at him oddly, as if he were unusual, when he thought that she was the peculiar one because she did not have one. There is vast ignorance about number forms because the topic does not ordinarily arise: people do not speak of what they do with their number forms, nor do they show them to other people. Number forms are like sex, though less exciting.

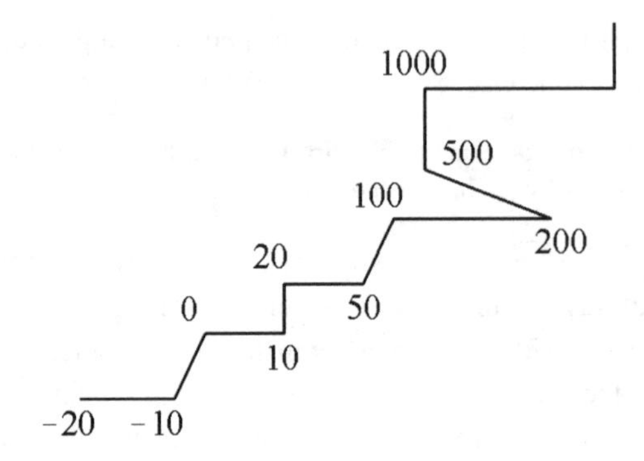

FIGURE 6 A form with negative numbers.

The forms are usually two-dimensional, though some peoples' twist through space. Most forms are wire-like, though some have width. Sometimes they have regions of varying brightness, or of different colors. Number forms come in great variety. Sometimes they include negative numbers as well as positive ones, as in Figure 6. They can be connected or, as in Figure 7, disconnected.

By now, you should know if you have a number form or not. I know that I do not: if asked to think of 17 I see nothing in my mind, and if someone asked me to think really hard about 17, the best I could do would be to see the digits 1 and 7 in black on a gray background. How very dull. How much more fun it would be to have the number form in Figure 8, one not just with shape, but color too! The colors were, according to the person with that form,

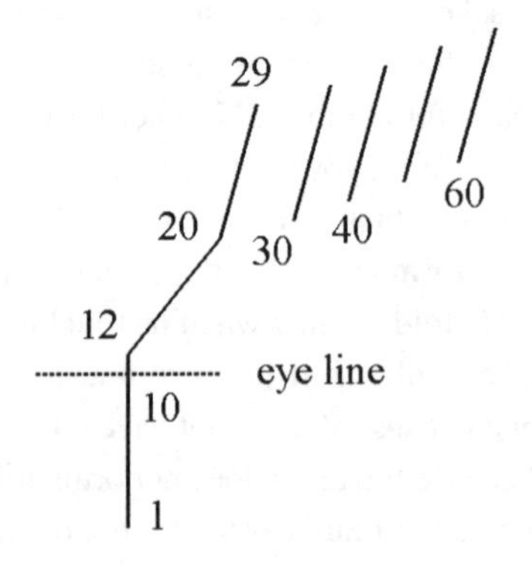

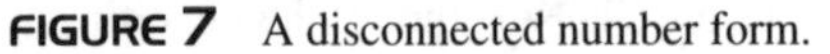
FIGURE 7 A disconnected number form.

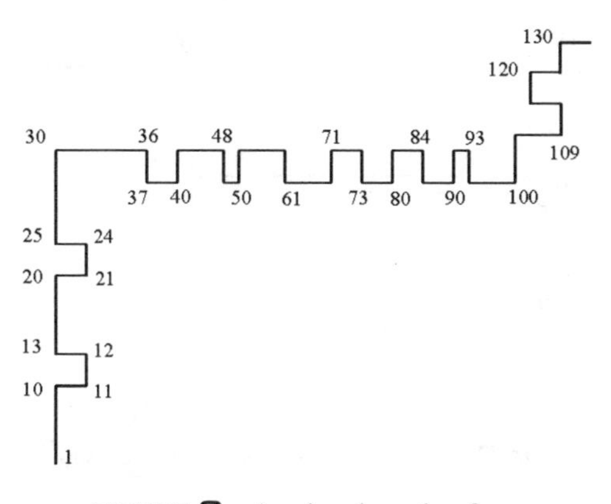

FIGURE 8 A colored number form.

0	scarlet	9	green	8	white
7	yellow-green	6	green	5	flesh color
4	dark brown	3	pink	2	blue
1	bronze				

Compound numbers have the tints of their components.

Number forms have not been studied much. As far as I know, the first person to take notice of them was Sir Francis Galton (1822–1911) in the 1870s:

> I have lately been occupied in eliciting the degree and manner in which different persons possess the power of seeing images in the mind's eye.... The subject bears in many ways upon psychological and ethnological studies.... The various ways in which numerals are visualized is but a small subject, nevertheless it is one that is curious and complete in itself. My data in respect to it are already sufficiently numerous to be worth recording, and they will serve to show that parallel results admit of being arrived at in other directions. [1]

He noted that

> the power of seeing vivid images in the mind's eye has little connection with high or low ability or any other obvious characteristic.... It is not possessed by all artists, nor by all mathematicians, nor by all mechanics, nor by all men of science. It is certainly not possessed by all metaphysicians, who are too apt to put forward generalizations based solely on the experiences of their own special ways of thinking, in total disregard of the fact that the mental operations of other men may be conducted in very different ways to their own.

He gave many examples. They show that number forms were the same in the nineteenth century as they are now. One of them is Figure 9. The possessor of that form said that

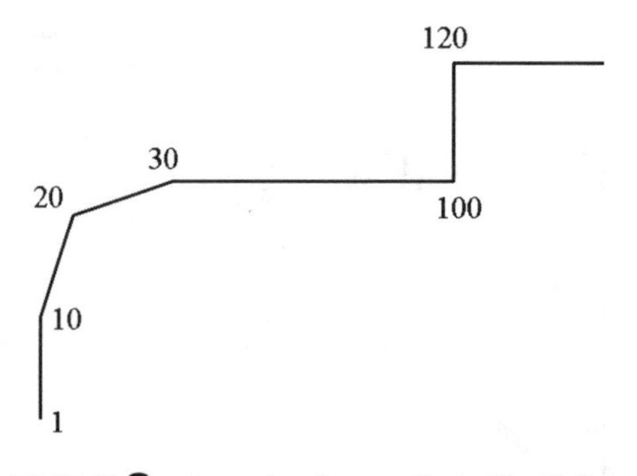

FIGURE 9 A number form collected by Galton.

> figures present themselves to me in lines (as in the picture). They are about a quarter of an inch in length, and of ordinary type. They are black on a white ground. 200 generally takes the place of 100 and obliterates it. There is no light or shade, and the picture is invariable.

That is similar to what possessors of number forms say today.

Galton's purpose seems to have been to support his opinion that the reports of people on the workings of their minds were reliable, something that evidently was not generally agreed on at the time:

> Therefore, although philosophers may have written to show the impossibility of our discovering what goes on in the minds of others, I maintain an opposite opinion. I do not see why the report of a person upon his own mind should not be as intelligible and trustworthy as that of a traveller upon a new country, whose landscapes and inhabitants are of a different type to any which we ourselves have seen. It appears to me that inquiries into the mental constitution of other people is a most fertile field...

In his *Inquiries Into Human Faculty*, Galton said that additional data had enabled him to arrive at some conclusions about number forms.

> The pattern or "Form" in which the numerals are seen is by no means the same in different persons, but assumes the most grotesque variety of shapes, which run in all sorts of angles, bends, curves, and zigzags.... The drawings, however, fail in giving the idea of their apparent size to those who see them; they usually occupy a wider range than the mental eye can take in at a single glance, and compel it to wander. Sometimes they are nearly panoramic.
>
> The Forms have for the most part certain characteristics in common. They are stated in all cases to have been in existence so far as the earlier numbers in the Form are concerned, as long back as the memory extends; they come into view quite independently of the will, and their shape and position, at all events in the mental field of view, is nearly invariable....

> The peculiarity in question is found, speaking very roughly, in about 1 out of 30 adult males or 15 females....
>
> It is beyond dispute that these forms originate at an early age; they are subsequently often developed in boyhood and youth so as to include the higher numbers, and, among mathematical students, the negative values.... Not one in ten is able to suggest any clue as to their origin. They cannot be due to anything written or printed, because they do not simulate what is found in ordinary writings or books. [2]

He also concluded that heredity was involved in number forms. I can believe that, since among my collection of forms are two possessed by identical twins. Their forms, however, were not identical. Galton hypothesized that among children there was a continuum of forms, from none at all, through vague ones, to clear and fully developed forms, and that over the years the vague forms tend to disappear while the fully developed ones remain and even develop further. This, Galton said, explains the sharp division among adults between those who have forms and those who have none.

His book contained 63 examples of forms, some ordinary (Figure 10) and some picturesque (Figure 11). Galton's work stimulated interest in the subject. In 1893, G. T. W. Patrick, Professor of Philosophy in the State University of Iowa, wrote to give

> ... some explanatory remarks and a few suggestions toward a future theory. It is hoped, too, that further attention may be called to the subject, and other contributions made to this curious chapter in psychology. [4]

He asked a group of seventy-five students if they had number forms and found that four did.

> This would correspond roughly with Galton's estimate that one out of every thirty adult males, and one out of every fifteen adult females, has a number form. My own later experience, however, has developed the fact that such a mode of investigation does not discover the full number of persons possessing forms, simple or complex. There are several reasons for this. The subject is not commonly understood when first presented. It would seem that a person

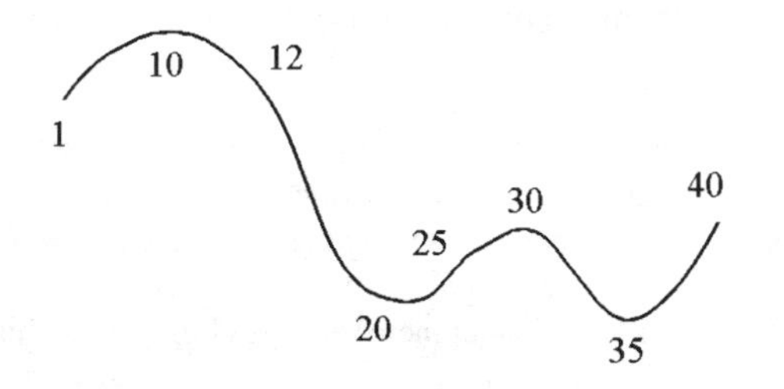

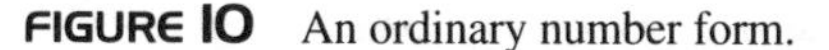
FIGURE 10 An ordinary number form.

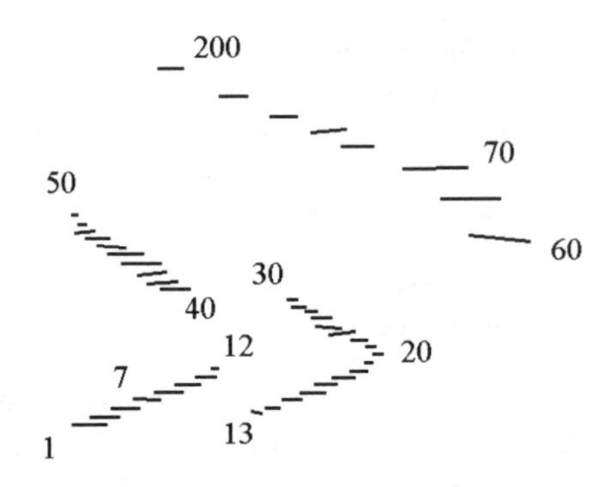

FIGURE 11 A picturesque number form.

> having even a complicated form might live and die without knowing it, or at least without once fixing his attention upon it or speaking of it to his nearest friends, although such a one might use his form in daily computation. It seems to him quite natural to see the numbers in that way, and the thought may never enter his mind that others should see them differently. Again, if one is conscious of a peculiar form, he regards it as an idiosyncrasy and exhibits a certain shyness in revealing it. For this reason it is especially hard to get all the number forms from a company of children. They do not like to be laughed at, and will willingly keep silent about anything which they suspect may be another of those idiosyncracies causing such mental torment to many children. Finally, those who do not have complicated forms are apt to think that the little curve, twist, or angle in which they see the numbers is quite too trifling a matter to mention. I am inclined to believe that one out of six adults would be a more accurate proportion, that the proportion among children would be still greater, and that is perhaps a little more common among women than men.

I am more inclined to agree with Galton that the proportion is 5% or less. It is of course possible that the proportion changes with time. Television may have caused many number forms to wither.

Patrick confirmed Galton's findings that forms are permanent and invariable, that people with forms cannot think of numbers without seeing them, and that some people use them in computations while for others they are just *there*. He suggested a cause:

> Speaking very generally, however, their origin may be traced to one great cause—namely, the attempt or necessity of children to give a concrete form to the abstract. Now, numbers are among the first abstractions that children have to wrestle with. Our earliest abstract ideas, perhaps also our later ones, are, as it is now well known, either mere samples of individual things, or else a composite picture of them. The child's concept of boy, girl, dog, horse, are nothing more than visual pictures of some particular boy, girl, dog, horse, or

> else a composite picture of a limited number of individuals. Now, numbers do not admit of such composite pictures. They are bald abstractions that the child must manage in some way.

Some children manage them with number forms.

> A number form thus becomes a little system of topical mnemonics. Its continuance, either in the individual or, in the cases of inherited forms, in the family, is of course due to physiological conditions.

Professor Patrick was doubtless correct. That babies are not born with number forms already in their heads is shown by how the forms almost all follow the decimal system, with bends and curves at 10, 20, 100, and so on. Babies know nothing of counting by tens; that has to be learned, and the number forms cannot predate that learning. Number forms must form when children learn to count, and the process of their formation is forgotten along with all of the other things that we forget about being 3, 4, or 5 years old.

Evidence for this is contained in [3], an article that appeared in 1893. Its author, Adelia Hornbrook, a teacher at the Classical School of Evansville, Indiana, was the possessor of a number form, one which she invariably used when doing computations. She reached the conclusion that everyone with forms used them similarly:

> Investigations among the few people whom I have met who had number forms, and hence were responsive to questions about them, show that the forms, however complicated and irregular they may be, are always a great assistance in arithmetical calculations.
>
> Being asked if they can work more quickly and accurately with than without their number forms, they invariably reply, in substance: "I don't know. I have no idea how I should work without them, for I always use them."

Her classroom experience confirmed this:

> One of my pupils showed great readiness in dealing with fractions when first introduced to the subject. The question "What is 9/7 of 21?" being asked, she gave the correct answer instantly, while the rest of the class were getting ready to solve it. "How did you get it so quickly?" asked the teacher. "Why," said the girl, "there is 21, and the other two 3's that go on and look different." She could not make her meaning clear, and she was not urged. At another time being asked for 7/8 of 100 she instantly gave 87 1/2. Being invited to explain her quick method to the class, she replied: "I cut 12 1/2 off from the right hand side of the 100," making an illustrative gesture with her right hand.

Since number forms are so useful, she thought, Ms. Hornbrook conceived the idea of trying to *implant* them in her students. She tried for six years.

> Limits of space forbid the mention of all the devices by which we have tried to fix the forms in the minds of the children.

She concluded

> Whether they have permanent number forms or not we cannot yet tell, but they certainly had them while they were learning the multiplication table.

That sounds like wishful thinking. I am not convinced that she succeeded in creating a number form in a mind that did not have one already. Some children become skilled in saying what they think a teacher wants to hear, and that may be part of the explanation for her conclusion.

Number forms, I think, can neither be created nor destroyed. As Ms. Hornbrook said herself,

> As an indication of the tenacity with which the mind clings to its own number form, I may say that with all my thought and work upon the form given to the children I never use it in my own calculations, seeing numbers and reckoning with them by means of the inverted and imperfect number form which I have had since I can remember.

Interest in number forms has died down in the last hundred years. Although our understanding of how the brain works has advanced since 1880, it probably has not advanced enough to deal with number forms. Another hundred years or so may be needed.

In any event, if you have a number form, be grateful. Think how much better off you are than the 97% of the population without them, to whom numbers are at best blurs and in any case not worth looking at. Treasure it. Make a copy of it and send me one, even if it not as impressive as the examples in the pictures. I wish *I* had a number form.

References

1. Galton, Francis, Visualised numerals, *Nature*, **21** (1879–80), 252–256.
2. ———, *Inquiries into Human Faculty*, Macmillan, London, 1883.
3. Hornbrook, Adelia R., The pedagogical value of number forms, *Educational Review*, May, 1893, 469–480.
4. Patrick, G. T. W., Number forms, *Popular Science Monthly*, **42** (1892–93), 504–514.

CHAPTER 19

Mrs. L. Dow Balliett

We have seen how the ancient Pythagoreans, with their idea that all is number, originated number mysticism. The Neopythagoreans, more than half a millennium later, extended and deepened the mystical properties of the integers. During the middle ages, number mysticism and symbolism simmered along in Europe, but with no great advances being made. The revolution that created modern numerology had to wait until the late nineteenth century, in Atlantic City, New Jersey, where Josephine (Dennis) Balliett carried it out, as far as I know singlehandedly.

The revolution consisted of attaching numbers to people, and then asserting that the mystical characteristics of the number were likewise attached. No one had had this idea before. For the Pythagoreans, original and Neo-, number mysticism was a medium through which to view the universe, so as to understand it better. Other than the efforts to attach the number of the beast to people, medieval number symbolists had much the same idea as the Pythagoreans, using numbers to enhance and embellish, but not applying them to individuals.

In retrospect, the revolution may not seem all that revolutionary, since astrologers had been attaching astrological signs to people for centuries and applying their mystical properties to make predictions and give advice. But this is hindsight, which clearly shows that everything that has happened was inevitable. Every theorem is trivial after it is proved. Mrs. Balliett (the wife of Lorenzo Dow Balliett, physician) had a new idea, and deserves all the credit that idea deserves.

My identification of Mrs. B. as the sole originator of modern numerology should not be taken as authoritative. Numerology scholarship is an undeveloped field. I *think* that Mrs. B. was the first, but that is only because I have not been able to locate any predecessor. I have searched as diligently as I know how to, but I may have missed something. If anyone knows of a previous modern numerologist, I would be very pleased to hear about him or her.

I was not able to find out very much about Mrs. B. either. She was born in 1847; where, I do not know. Lorenzo D. Balliett first appears in the City Directory of Atlantic City in the 1894 edition. In 1899 there is a separate entry for Mrs. L. Dow Balliett, principal. (She was principal of the School of Psychology and Physical Culture, Atlantic City, about which I have no other information.) After 1904 the entry reads L. Dow Balliett (wife Josie D.), and in 1928 it changes to Josephine Balliett, widow of L. Dow Balliett. There is no entry in the 1930 or later editions, so it seems likely that she died, aged 82 or 83, in 1929. The Library of Congress has some of her works listed as by Sarah Joanna Balliett, so she may have changed her name, perhaps for numerological reasons.

She was a prolific author. Among her works, many of which had several editions, are

The Body Beautiful
Musical Vibrations of Speaking Voice
Philosophy of Numbers, Their Tone and Colors
Beyond Sight
Nature's Symphony, or Lessons in Number Vibration
The Day of Wisdom According to Number Vibration
Universal Music.

One of her books has her picture ("reproduced from *First Citizens of New Jersey*"), showing a woman with upswept hair and a fine-featured oval face, full of intelligence. She was a co-editor of *Early History of Atlantic County, New Jersey* (1915), and was evidently a person of energy and attainments.

Let us examine one of her popular books [1], whose title page reads:

How to Attain
Success Through
The Strength of
Vibration
A System
of Numbers
as Taught by
Pythagoras

It begins:

> In sending out this work, the author feels joy in being able to give a fundamental from which many of life's problems may be solved. It is founded upon mathematical principles in the same manner as music is developed.

> Words are analyzed to find their exact place and meaning. There is no guess work to be found in this book concerning the gems, fruits, etc., to which you vibrate; they are all worked out from the one theory of vibration found in name and birth number. If one part is true, all is true. In this way you can find with ease things related to you, which seem mysteriously hidden.
>
> It is the author's earnest wish that you will not stop at a few things recorded as vibrating to your especial numbers, but that you will bring to outward consciousness all things in your environment and thus be enabled to solve many unknown problems. [1, p. vii]

A difficulty that arises is that "vibration" is nowhere defined, nor the mechanism that causes a person to vibrate to specific gems, fruits, and so on. I suppose that we cannot cavil over this any more than a student of geometry can complain that point, line, and plane are nowhere defined, nor the mechanism that produces them. *Everybody* knows what points, lines, and planes are. If you are so unfortunate as to not have any understanding of the meaning of "point," "line," and "plane," then geometry, and geometry books, are going to be meaningless to you. It is, I think, the same with vibration. To get the full benefit from books of numerology, you must *know* what vibration is. You must know about it before you start reading the book, just as you must know about lines and planes before starting to read a geometry text. And, just as your knowledge of lines and planes came from a source other than a geometry book, your knowledge of vibration must have been derived from somewhere other than a numerology book before you can make sense of numerology.

Though I have never read a numerologist claiming that numerology and geometry have equal stature, a case could be made. Are we justified in sneering at those who write knowingly of vibrations, and dismiss them as writing nonsense about something that does not exist? If so, how would we defend ourselves against the sneers—not that there are any, such is the prestige of mathematics—about studying undefinable and nonexistent lines and planes? Is there a difference?

If you wanted to, you could maintain that there is not. Since both "line" and "vibration" are undefined, they may take on meanings that differ from person to person. Though we may be able to talk together about lines pleasantly and with no disagreements, what is in my head when I think "line" may be quite different from what is in yours. I have no way of knowing, and can have no way of knowing. The mystical inwardness of "line" can be perceived only wordlessly. Similarly, no one can explain "vibration" to you in words. If you lack that wordless insight into its true nature, then you may say, "Why, vibration is nonsense, it doesn't exist," but what you are really saying is, "I am incomplete, I have a blank where I should have my knowledge of vibration."

The point is not that geometry and numerology are all of a piece, six of one and half a dozen of the other, equally valid, but that "Nonsense!" is an insufficient response to the claims of numerologists. We have to do better than that.

But we can do better. Regardless of our mystical appreciation of lines or vibrations, geometry differs from numerology because it has *applications*. Geometry *works*. Numerology does not. Numerologists would like us to think that it does, but it does not. Mysticism operates in one direction only—inwards. There is nothing wrong with this, since it is better to have a rich and vivid inner life than a barren one, but claims that mystical insights can lead to power, influence, or increased net worth can properly be met with "Nonsense!" It is up to the numerologists to provide evidence to the contrary, and this they have not done, nor can they.

Since Mrs. B's work was the foundation of modern numerology and a model for those who have come after her, it is worth a close look—if for no other reason than that after looking at it closely, you will not have to look at any other numerological work closely, or at all.

Chapter 1 is titled "The Principles of Vibration." Principles, that's what we want, we think. Theorems follow from principles. This is what we get.

> Ever since the middle of the VI Century, B.C., when Pythagoras gave to the ancient Greeks his system of numbers, each succeeding generation has been attracted, more or less, to his theory, principally through his grand conception of the Music of the Spheres
>
> All the way down the ages each generation has regarded the idea as interesting but fanciful. There are some who know it is true.
>
> As we understand the teaching of the old master, all things are in a vibratory condition; the higher the vibration the more spirit force an object contains and the more positive it is in its nature; the slower the rate of vibration, the less force it contains and the more negative it is in its action.
>
> [1, p. 9]

High notes are good, the higher the better. That high notes can be shrill and screeching, and low notes sometimes soothing, the author does not consider. The Pythagoreans knew that the planets pursued their courses through the heavens at roughly constant rates, and their periods of revolution had fixed ratios. Ratios implied sound, 2:1 the octave, 3:2 the fifth, so the planets were humming as they moved—the inaudible (and mystical) music of the spheres.

> Everything, from a grain of sand to a man, is vibrating at its own rate and round its own centre which is its keynote. As each object seeks its own, so everything, both great and small, has found its place in the great chorus of nature, God's choir. [1, p. 9]

Mrs. B. had said earlier that "the Bible, as well as the works of Pythagoras, Plato and the other philosophers, has been used to develop the system." Pythagoras has become an honorary Christian, his teachings, or rather Mrs. B.'s version of them, harmonizing with no difficulty with the Bible.

Mrs. B. was ready for the "ordinary man" who could not feel the vibrations:

> Men have a normal gamut of sight and hearing; highly endowed individuals having a more extended gamut and still a few others, especially developed, see colors and hear sounds that the ordinary man says do not exist. But the world's unbelief will not convince the endowed man that he does not see and hear beyond the ordinary; and the fact that our ears are not attuned to it would not have made Pythagoras believe the music was not all around us and that each was not playing his part in the world's great chorus.
>
> [1, pp. 9–10]

All is number, and if numbers vibrate, then all is vibration. And those who cannot feel the vibration are disabled and imperfect. This does not help much in grasping the principles of vibration.

Next comes some dubious history:

> This thought of playing a part in Nature's Chorus was a delectable one to the Greeks and before the short Earth life of Pythagoras was ended many of them worshipped him as a god, and schools and universities headed by the greatest philosophers of the time were started to teach his system; but none of his followers were sufficiently endowed to develop the teaching further than he had given it. Nevertheless, each succeeding generation has put out a faint hope that some day the "Music of the Spheres" would be proved a reality to the masses. [1, p. 10]

The world of Pythagoras is too far away and too alien for us to understand anyway. Everything was new. The idea that reason could be used to get at truth had just entered the head of the race. There were no explanations for anything. *Anything* could be true. The planets *could* be vibrating. All *could* be number.

It is impossible to get inside the heads of the Pythagoreans. They were human, even as we, but when they looked at the world they saw something utterly unlike what we see. Societies change: people in the eighteenth century would react with disgust and horror if they had been shown a picture of what we think is the glorious sublimity of the Grand Canyon. People in the nineteenth century *stood* differently than we do, as many pictures show. We are stuck in our own time and though we may struggle to get out, we cannot succeed. We will never know what, or how, Pythagoras thought. To assert that the ancient Greeks thought delectable thoughts about singing in nature's chorus is as silly as it is meaningless.

Since Pythagoras approached semidivinity, he could say no wrong:

> Pythagoras was so highly endowed that he was a freed soul, and because less highly developed men have not reached the heights to which he attained, they have no right to call his teaching false. [1, p. 10]

Though Mrs. B. could hardly come out in support of the Pythagorean reincarnation of souls, since that is very hard to reconcile with either the Old or New Testament, she nevertheless believed in conscious *pre*incarnation:

> No one is out of place. Our highest part chose the path in life by which we entered the world. The individualized spirit, called the Soul of Man, knew what experiences it needed when it took upon itself the mystical substance called the body, in order to perfect itself in the Christ life, and chose the vibration which would draw to itself the experiences needful for his highest spiritual growth. Thus he chose the rate of vibration best suited to his need, and in order that the intellect might comprehend it, attached the intellectual form of numbers.
>
> It must now be plain that these numbers which we are going to find are not an accidental gift, but were decided upon by our highest spiritual nature. [1, p. 11]

The implication is that parents do not have free will in assigning names to their children. They think that they do, but we pick our own names. Or at least, we make sure that our names will have the right number and hence the right vibration.

Now we get down to numbers:

> Pythagoras says that every letter of the alphabet has its own rate of vibration and color. He divided numbers in this way:

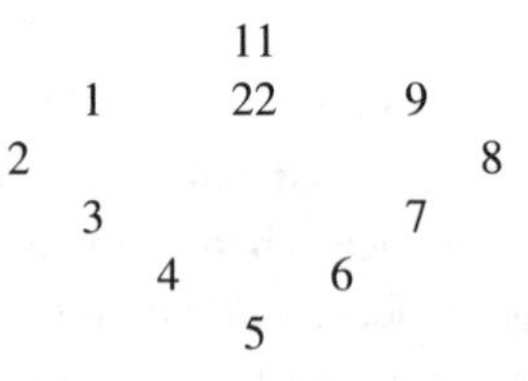

> into odd and even; into limited and unlimited, and gave the preference to odd numbers. But we consider all of them necessary. [1, p. 11]

There is no evidence at all that Pythagoras ever made any correspondence between letters and colors, and crediting him with being the founder of gematria is likewise unsupported.

Mrs. B. says

> This system of numbers is founded upon Pythagoras' Ten Fundamental Laws of Opposites, which should be carefully studied, as they are the key. [1, p. 12]

Mrs. B. does not say the key to what, nor does she give the Laws until page 60, but there they have the whole page to themselves:

1. Limited and Unlimited
2. Odd and Even
3. One and Many
4. Right and Left
5. Masculine and Feminine
6. Rest and Motion
7. Crooked and Straight
8. Light and Darkness
9. Good and Evil
10. Limited and Unlimited

You may notice a similarity between the first and last pairs. All other authorities give the last as

10. Square and Oblong

You can see why Mrs. B. made the error, either on purpose or by accident. "Square and oblong" just does not have the same scope and universality as "good and evil," "one and many," or any of the others. "Hot and cold" would have been much better, as would "smooth and rough," "spiritual and material," or even "upright and prone." "Square and oblong" sounds trivial in comparison. Even "prime and composite" would have been better.

We are still in the chapter "The Principles of Vibration."

> Individual lives, as we see them, move as do the pendulums of clocks. Some sway with broad, full sweeps from one extreme side to the other, while others move in a more limited compass. All these clocks fulfill their mission as time-tellers. Developed souls vibrate to strong numbers, and their deeds are marked by the world as good or bad. The smaller sweeps are less noticed, yet they may be making strong vibrations, which will be expressed by different names and numbers. [1, p. 12]

See how the march of technology can have effects bad as well as good. Pendulum clocks are no longer much seen, so Mrs. B.'s image would not occur to writers today. Digital watches can give no similar inspiration.

> There are as many periods in one's life as there are parts in the baptismal name, including the Mother's name before marriage. [1, p. 12]

This is not entirely clear. Presumably it means that if you have a first, middle, and last name, then your life will have four periods. Or perhaps more if the mother's full name is meant.

> Read each part of the name separately; find its value according to the letters in the name. Then find the sum of the figures comprising each name and the digit. Read each period in regular order, according to the vibration of each name. The digit of the whole name is the most important, as it shows the quality of vibration the person was born with, the name now used, as how the world regards him at the present time.
>
> The digit of birth path shows what part in the great chorus of life they come to take. It can not be changed. It must be met. It can be made harmonious or its opposite.
>
> Every life has its own song. That it should unconsciously sing all the way through life—it does this so long as there exists in the individual a desire to reach its ideals. Obstacles may be placed in its pathway, but so long as the higher leads, the song goes on and mingles with the melody of all things that are vibrating their higher nature. It is this that clears the atmosphere of inharmony and attracts spiritual and material gifts. These gifts can not be separated, as the individual's holiest vibrations has called them. Life is then a song of joy and burdens disappear. [1, pp. 12–13]

Would that it were so! Here is how to make it so:

> To do this, find your birth vibration. This will show the part your higher nature wants developed. Accept its experiences as friends; when met and harmonized you free a glad note that is added to your infinite song of life. [1, p. 13]

Then she proceeds to give the letter-number chart that numerologists use, in spite of its non-Pythagorean character (Table 1).

1	2	3	4	5	6	7	8	9
A	B	C	D	E	F	G	H	I
J	K	L	M	N	O	P	Q	R
S	T	U	V	W	X	Y	Z	

TABLE 1 Letters to numbers.

She then shows how to reduce numbers modulo 9, by adding digits. Mathematicians take as their system of residues modulo 9 {0, 1, 2, 3, 4, 5, 6, 7, 8} but numerologists prefer {1, 2, 3, 4, 5, 6, 7, 8, 9}. This is quite Pythagorean, since the ancient Greeks had no digit 0, and it is superior numerologically, since no one would want to have the number 0. Zero is so lacking, almost negative. Nine is much better.

The idea of giving numbers to names in this way is very similar to the pythmen and pythmenes that we saw in chapter 4. How Mrs. B. found out about this ancient method of gematria (if she did) I do not know, and she does not tell.

Mrs. B., being musical, also assigned a key to each number:

c	d	e	f	g	a	b
1	2	3	4	5	6	7
8	9					

This is new, and non-Pythagorean.

In chapter II, "Numbers in Detail," Mrs. B. gets down to the character of numbers, a mixture of ideas—mostly Greek (Pythagorean and Empedoclean), but Hebrew and Christian as well:

> The first 3 numbers of 1, 2 and 3 compose the sacred cycle of creation, assimilation and expression. They also denote the action of the mind found in impression, re-impression and expression. When this trinity of numbers is found in digit of single name or birth vibration it shows the power of harmonious expression. The cube is formed of numbers 4, 5, 6, 7.
>
> [1, p. 20]

I have no idea what that means, or how to get a cube out of 4, 5, 6, and 7. Cubes have six faces, eight vertices, and twelve edges, but there does not seem to be a 22 ($= 4 + 5 + 6 + 7$) there. Mrs. B. was probably drawing on some uncredited source. It would be helpful to know what it was.

> These are the vibrations that seem limited in action, and yet each contains enough success for the happiness of any one bearing them if they seek the highest of their vibration. The unlimited numbers of freedom are 8, 9, and 11. To these is added 22, which is closely related to 11. Pythagoras considered 1 and 22 as numbers possessing a mystical character, as they are the beginning and end of the Hebrew alphabet. [1, pp. 20–21]

Where Mrs. B. got the idea that Pythagoras was acquainted with the Hebrew alphabet is another mystery. Perhaps it was from her unknown source, though it could have been pure assumption—highly developed souls range widely, and know much.

> 11 being the highest point of its vibration, 22 possesses the character of 2 with added strength and freedom. The four elements of Earth, Air, Water and Fire explain broadly the character of number vibration.
> Earth vibrates to 7—it contains all the highest in limited vibration.
> Air vibrates to 1—Is found in all places and in all things, but is unseen.
> Water vibrates to 22—Whose strength lies in cohesion; its power lies dormant when separated from the whole.
> Fire vibrates to 11—God spoke through fire. So is 11 a high priest, whose life must deliver some kind of a sermon. [1, p. 21]

Next comes Mrs. B.'s revolution. Past number mystics meditated on the mysteries of number and derived therefrom insights into the character of numbers. Since 4 is a perfect square, it is solid, dependable, slightly dull. Even numbers are easily divided into two, hence weak, hence female. The properties of the digits are all very understandable, mystically, and they were preserved for a thousand years by the Pythagoreans, the Neopythagoreans, and those who vibrated to them. Mrs. B. made the giant step of applying the characteristics of the numbers to the people who, in one way or another, bore them. *Applied* mysticism! A revolutionary idea! Revolutionary ideas can seem simple, after someone else has had them. Those who have them are the geniuses of the race. For those of us who think that mystical insights have no applications and to talk about them is pernicious nonsense, Mrs. B. was a misguided genius, but she was a genius nevertheless. The present popularity of Balliettan numerology proves it.

Here is what a "1" person is like:

> No. 1
>
> indicates a strong nature in which the power lies to create, develop and govern all things pertaining to this life; but unlike the other numbers it requires a complete overcoming of self before it attains its highest success. It makes all the other vibrations active rather than acting as an individual principle.
>
> The perpendicular line of 1 signifies Truth and Separateness. A No. 1 person will mingle with the world, but is never really one of them. Persons vibrating to No. 1 must learn through their own experience, that everything gained, until self is overcome, comes through affliction. Friends will leave them or so abuse their confidence that they would have cause to separate; but they will not do so when they realize that to love better is their privilege and to be able to love without thought of return is what raises the No. 1 individual and gives him such happiness as a weaker nature can not comprehend. They are original and comprehensive in thought and always expect their opinions to be respected.
>
> They stand for the points in Geometry, seeing deeply beneath the surface.
>
> Pythagoras gives them the freedom of limited and unlimited opposites.
>
> No. 1 is the creative power, the line between Heaven and Earth.
>
> The mineral is copper, but when they have passed through the period of sacrifice of self, they draw to themselves some iron, which makes the strongest foundation.
>
> The color is Flame. [1, pp. 21–22]

The number of *copper* is $3 + 6 + 7 + 7 + 5 + 9 = 37 \equiv 1 \pmod 9$, and of flame is $6 + 3 + 1 + 4 + 5 = 19 \equiv 1 \pmod 9$. So it is not only people who have numbers, but substances as well. Anything that has a name has a number, anything that has a name vibrates, the universe is filled with vibration, the music of the spheres!

A difficulty with this mystic vision is that a few pages ago, Mrs. B. was telling us that unborn souls choose their names so as to have the proper number.

If copper is to have a number, we either must believe in the Spirit of Copper and become complete animists or somehow work out how it was that as the language evolved, names of metals and other objects were assigned to be consonant with their vibrations.

Though there are echoes of the Neopythagorean properties of the monad in that description of 1, they are not terribly strong. Mrs. B.'s description of 2 bears a much closer resemblance to the old properties of the dyad. And, of course, the properties are applied to people.

> No. 2
>
> No. 2 has the mother nature and is intuitive. Persons vibrating to No. 2 have temperaments so fine that they are not fitted to stand alone in hard places as No. 1. The ore that predominates in their compositions is gold—which is their color,

[gold: $7 + 6 + 3 + 4 = 20 \equiv 2 \pmod 9$.]

> and they usually attract their own if they receive and follow the lead of other minds and from them select their ideals.
>
> They water and nourish the seed others plant, and they themselves usually reap the harvest.
>
> They are closely connected with No. 1 and know how to deal with things in a material way. When these people give a truth to the world they know how to clothe it in beautiful language.
>
> They make good lawyers, seldom reaching the highest places, but like Hanna, they mould the minds of the people and are great leaders.
>
> They are also like John, who leaned upon the Master's breast, and Sarah, who gave the world an Isaac.

[John: $1 + 6 + 8 + 5 = 20 \equiv 2 \pmod 9$; Sarah: $1 + 1 + 9 + 1 + 8 = 20 \equiv 2 \pmod 9$.]

> No. 2 is the conjunction of Spirit and Matter. They are peace-makers. Great arbitrators are found under this number. They are seldom really great men but draw great men to them and it is necessary for the success of a great man that he have a No. 2 associated with him. Their color is gold and when they reach its vibration, which means silence and peace, they attract the mineral gold to them. In Geometry 2 is lines—they see straight ahead and do not waste their force trying different directions as does
>
> No. 3,
>
> which is the outward expression of the Christ principle of the Trinity.
>
> [1, pp. 22–23]

Mrs. B.'s characteristics of 2 could have been written by Iamblichus. With the difference—I am repeating myself, but I do not want Mrs. B.'s new idea to be missed—that they are applied to people whose names vibrate to 2.

Since she was moving ahead to people whose names vibrate to 3, let us follow:

> It is their mission to follow the example of the Christ—to heal the sick and bless the world.
>
> Others should build homes for them, but No. 3 should bless them.
>
> In Geometry it means surface and these people can see a long distance and behold large spaces of Earth which they wish to see a happy meeting place for individual souls.
>
> Theirs seems like a borrowed life lent for a purpose like the Christ life. They gather the blossoms No. 1 planted and rejoice over the happiness they give.
>
> They are inclined to waste their energy and are not steady workers with their hands.
>
> No. 3 is the vibration of expression wherever found. Most musicians, artists and actors vibrate to one of the threes. It is they who can interpret and bring forth the silent hidden voices of all things. They brush from closed eyes earth's dust that the revealed glory of the Divine may be seen. Especially is this true if the digit of separate name is 1, 2, 3, or the digit of the whole compose these numbers.
>
> They form the sacred Trinity. Within their vibration potentially present is creation, assimilation and expression. If found in birth vibration as digit of month, day, year, or digit of whole, their unity should be regarded, as original development should follow. When a 3 vibrates to the material part of vibration 2 it can easily express the character of a Judas as that of Mary, the perfect expression of divine womanhood. It should vibrate to 1 or 2.
>
> They vibrate to no special color but use the color that presents itself. To be happy and make others so is their mission.
>
> They are gleaners.
>
> The law of opposites is "One and Many." [1, pp. 23–25]

After the characteristics of numbers are understood, we must be sure we know what our number is. For Mrs. B. the name given at birth is not destiny, since names can change:

> The name of a man is exactly as he writes it, either with initials or name entirely written out, must be taken to find what the man is now doing—the plane on which he is living. The digit of all the baptismal names added to the digit of the mother's maiden name

[Still not entirely clear—*all* the mother's maiden name, or just the surname?]

> shows the quality of the individual, what he is capable of and what is being held in abeyance.
>
> At certain times in life we use one name or initial, afterward using another. We are always to the world exactly what the name or initials we are using vibrate to.
>
> A pet name given to you by any one, shows what you stand for to that person. [1, pp. 37–38]

Now we can proceed to applications. When choosing where to work,

> One should be as careful in business in selecting his street and number as in choosing a wife.
>
> If one possesses one of the strong numbers—8, 9, 11 or 22, he should go to a place that vibrates to one of these numbers.
>
> For instance, one possessing 8 or 9 should seek a place like Boston, as it vibrates to 22.
>
> If one does not possess a strong number he would better seek a city with a weak number.
>
> A person with a weak number might succeed in New York, which vibrates to 3. In Boston he would be overweighted.
>
> If John Hood—8, a strong number—wishes to make money he should go to Boston, which vibrates to 22.
>
> As the strong numbers attract each other, in seeking a home, he should go to Beacon street—22, the spiritualized temple. He could not succeed in a degrading business in Boston, but in any business that will benefit humanity he will succeed. Should he go on a street which has not a strong number, he should seek a house number that vibrates to 8, such as 17 or 134.
>
> [1, pp. 40–41]

There follows a sad short story, showing the effects of ignorance of numerology:

> Take the couple Barclay Rider Ramsey and Beatrice Emma Tilton. Barclay is 8, Rider 9, and Ramsey 9, giving him a name number of 8 plus 9 plus 9 equals 26 equals 8.
>
> He was born October 1, 1859. October is the 10th month and the digit of 1859 is 5—making his birth number 1 plus 1 plus 5 equals 7.
>
> Beatrice equals 5, Emma 5 and Tilton 9, 9 plus 5 plus 9 equals 5.
>
> Beatrice was born January 10, 1865. January is 1, the 10th day is 1, and the digit of 1865 is 2, making her birth path 1 plus 1 plus 2 equals 4. Now the two stand thus:
>
> Barclay Rider Ramsey, born October 1, 1859.
> Name number 8 plus 9 plus 9 equals 8.
> Birth number 1 plus 1 plus 5 equals 7.
> Beatrice Emma Tilton, born January 10, 1865.
> Name number 9 plus 5 plus 9 equals 5.
> Birth number 1 plus 1 plus 2 equals 4.
>
> By the law of attraction the vibration of Barclay being 8, attracted the woman Beatrice, whose first name also vibrated to one of the strong numbers.
>
> They first met on the strongest plane—the plane where the strongest feelings are developed for pleasure or pain.
>
> They married while Beatrice vibrated to the strong magnetic number 9, while Barclay vibrated to the strong number 8; 8 and 9 being in harmony, as they form the greater part of the divine Trinity of numbers. The 9 faculty of Beatrice fascinated her husband with the versatility of her personality while the vibration lasted.

[Alas! The vibration does not always last!]

> After a while, as the cycle of time turned the wheel, another vibration took the place of Barclay and Beatrice, 9 and 8 being in a manner held in abeyance. Rider and Emma came into action. Rider now vibrates to 9 and Beatrice to the Emma number of 5. In this condition the woman is the loser.
>
> The man is gaining in strength and ability, while the woman acting under the weaker number 5 usually dissipates her force. Had she known that 5 is in accord with 9 and allowed her husband to guide her, she might have led a peaceful and happy life. But she came down the path of life called 4 and threw herself into a turmoil of discord caused by the active principle of her name digit and the hard intellectual plane of 4. She frets and nags her husband because he does not vibrate with her upon the only plane she can now understand. 4 and 5 now urge her to arise with the lark and she is always busy and thinks others should also be always working. She longs to build a big house to be known to all men as her home. The man Rider, meantime, is vibrating in the higher atmosphere of rest and this annoys his wife. She fails to see that he is slowly winning a place of confidence and respect in the business and social world. The first vowel in Rider's name being i, he gains from the inner depths of his consciousness the knowledge he requires in his daily work. His birth vibration being 7, lower than his name number, indicates mental strength. It is easy for him to meet his duty day by day. All he needs is a harmonious environment to make his life comparatively easy. His wife, not understanding this, has great visions of the things he might accomplish by mere physical effort, while he feels he has left those things behind, his birth plane being 7.
>
> His wife coming down the physical, intellectual path of 4, can not understand that he has both knowledge and wisdom working for them, and he, having the closed number 7, feels that she is trying to pry into the sacred recesses of his being and resents it with silence.
>
> Finally, driven to desperation by her nagging, he is likely to seek forgetfulness in dissipation, although his soul revolts from the degradation; while she, grown weary and hopeless, may seek release in the divorce court from the man she still admires and whom she knows is in many ways her superior. If she but knew it they could always come together, on the vibration of their marriage name. If she will do this they may again live harmoniously, otherwise the vibration of peace and harmony will be lost. [1, pp. 46–49]

Days, being quickly reducible to numbers, vibrate as well. January 5, 1999, for example, vibrates to 7 since $1+5+1+9+9+9 = 34 \equiv 7 \pmod 9$. This has application:

> Persons with weak digestion will find that articles of food which have the same vibration as the day, will have a peculiar adaptation to their needs. [1, p. 57]

So, it is a day for spinach ($1+7+9+5+1+3+8 = 34 \equiv 7 \pmod 9$).

Numbers must be kept constantly in mind:

> Different articles of dress, such as shoes, hats, etc., vibrating at the same rate as their birth force, will give the feeling of fellow travellers.
>
> Many physical difficulties can be overcome by attention to this detail.
>
> [1, p. 57]

Life could become very complicated if you had to keep track of the number of all your clothes, all your foods, all your friends, everything that had a number, and correlate them with your number, the number of the day, the number of your town, all the numbers that belonged to you, so that they will vibrate harmoniously.

What Mrs. B. was doing was exploring the implications of her new idea. If numbers have applications, there is no reason why they should not be universally applicable. On page 67, we have this list:

THE STRONGEST AND WEAKEST PART OF YOUR BODY

No. 1 Throat, lungs, hands, nails, bones, epiglottis, bronchial, palate, bladder.

No. 2 Muscles and lips.

No. 3 Liver, glands, blood.

No. 4 Larynx.

No. 5 Mouth, pancreas, arteries.

No. 6 Kidneys, arms, cells, veins.

No. 7 Heart, legs, stomach, duodenum, ears.

No. 8 Spinal-cord, brain, skin, womb, back, spleen, nose.

No. 9 Eyes, head, feet, hair.

No. 11 Nerves.

No. 22 Cheeks, bowels, teeth.

On page 71, there is another application:

THE COMPOSER WHOSE MUSIC HAS A MESSAGE FOR YOU

No. 1. George Boehm.

No. 2. George Frederick Handel, Johann Sebastian Bach.

No. 3. Guiseppe Verdi, Mendelssohn, Schumann.

No. 4. Hayden, Wagner.

No. 5. Franz Liszt, Rubenstein.

No. 6. Schubert.

No. 7. Christopher Willibold, Brahms.

No. 8. Amadus Wolfgang, Mozart.

No. 9. Gluck

No. 11. Tschaikowsky.

No. 22. Joan Brahms, M. Von Weber.

All the names are exactly as given by Mrs. B., spelling and all, and have the numbers she assigned to them. Something odd was going on, or Mrs. B.'s arithmetic was slipping, since it seems that "Hayden": $8 + 1 + 7 + 4 + 5 + 5 = 30 \equiv 3 \pmod 9$ and "Wagner": $5 + 1 + 7 + 5 + 5 + 9 = 32 \equiv 5 \pmod 9$ are both misplaced. Even correcting "Hayden" to "Haydn" would not make the number right. "Schubert": $1 + 3 + 8 + 3 + 2 + 5 + 9 + 2 = 33 \equiv 6 \pmod 9$ however is where he belongs. Why Brahms appears twice, once as Joan, is not clear.

There is more. There is your bird, your flower, your month, and so on. Numbers can be assigned to anything with a name, numbers vibrate, and if you are wise you will vibrate in harmony.

Why modern numerology arose when it did and at no other time can only be speculated on. Mrs. B.'s time was a period of great interest in spiritualism and mysticism. Theosophy was founded by Helena P. Blavatsky in 1875, D. D. Hume was levitating in and out of windows, mediums and phrenology were flourishing. It was also a time of progress and belief in progress (Chicago's 1893 World's Fair celebrated "A Century of Progress") and in the ability of American know-how to solve problems (Mark Twain, *A Connecticut Yankee in King Arthur's Court*, 1889). It was a time when Mrs. B. could write

> *Question.* Then we really get what we deserve?
> *Answer.* Yes, it does not seem so, but the fault lies in us.
> *Question.* In what way does the fault lie in us?
> *Answer.* Nature is always true; she neither lies nor steals nor breaks promises. When we follow the law she fills all our needs; when we do not, she corrects us as we do our children.
> *Question.* Is Nature ever cruel or unkind?
> *Answer.* No, never. Man holds this vision, but Nature gives us for our use her world of things to use as materials for getting experience from which to develop our character into soul growth, and man abuses the gift when she allows him to fight, kill and destroy forms she will again make anew for other to use. She protects the just and allows man to punish himself.
> [2, pp. 38–39]

The time was ripe for numerology, as hindsight so clearly shows. It is still with us, and shows no sign of going away any time soon. As did Pythagoras, Mrs. L. Dow Balliett wrought well.

References

1. Balliett, Mrs. L. Dow, *How to Attain Success Through the Strength of Vibration*, 1905, reprinted by Sun Books, Santa Fe, New Mexico, 1983.
2. ———, *Number Vibration in Questions and Answers*, reprinted by Sun Books, Santa Fe, New Mexico, 1983.

CHAPTER 20
Numerology Books

Modern numerology is fairly uniform and one numerology book will, for the most part, be like another. They are like calculus books, though with slightly more variety. They will tell you how to find your number, or numbers—name number and birth-date number are standard, but there are also the life-path number, karma number, and others—give the properties of the numbers, and then proceed to fill the remaining space with examples, or go into how numbers interact (with $\binom{9}{2} = 36$ pairs to go through, this can consume quite a bit of space).

Immediately after the Balliett revolution, as after most revolutions, time was necessary before things settled down. In *The Power of Numbers* [2], a book with no date of publication but whose type and style give it the look of something originally published around 1920, to find the number of your name you do not just sum the values of its digits, you make a triangle (the author called it a pyramid but it is really a triangle), as

$$\begin{array}{ccccccccccccc}
\mathrm{M} & & \mathrm{I} & & \mathrm{C} & & \mathrm{H} & & \mathrm{A} & & \mathrm{E} & & \mathrm{L} \\
4 & & 9 & & 3 & & 8 & & 1 & & 5 & & 3 \\
 & 4 & & 3 & & 2 & & 9 & & 6 & & 8 & \\
 & & 7 & & 5 & & 2 & & 6 & & 5 & & \\
 & & & 3 & & 7 & & 8 & & 2 & & & \\
 & & & & 1 & & 6 & & 1 & & & & \\
 & & & & & 7 & & 7 & & & & & \\
 & & & & & & 5 & & & & & &
\end{array}$$

New rows come from old by addition modulo 9 or, if you are a numerologist, by casting out nines, which amounts to the same thing:

$$7 + 7 \equiv 5 \pmod 9 \quad \text{and} \quad 7 + 7 = 14;\ 1 + 4 = 5$$

have the same content. The last number, 5, is the author's number for Michael, different from the number that Mrs. Balliett and all other modern numerologists would give him:

$$4 + 9 + 3 + 8 + 1 + 5 + 3 = 33;\ 3 + 3 = 6.$$

Michael would have quite a different character, depending on which method was used. The properties of the integers in [2] are almost, but not quite, those that are now standard:

> *One* symbolises heat and light, wisdom and love. It stands for progress and improvement.
>
> *Two* symbolises procreation, fruition and the relation of all opposite things. It stands for doubt, hesitation and a lull in progress.
>
> *Three* symbolises intelligence, consciousness and resolution. It stands for movement, change and business.
>
> *Four* symbolises discernment, discretion and intelligence. It stands for work, reason and logic.
>
> *Five* symbolises judgment, discretion and intelligence. It stands for good fortune, plenty and surprises.
>
> *Six* symbolises peace, harmony, satisfaction and material well-being. It stands for co-ordination, marriage relationship and a spirit of understanding.
>
> *Seven* symbolises death, rest, wisdom and evolution. It stands for doubts, misunderstandings and darkness.
>
> *Eight* symbolises revolution, rupture, separation and expulsion. It stands for waywardness, loss and foolish actions.
>
> *Nine* symbolises regeneration, the birth of a new era, voyaging and extensions. It stands for powerful actions, unexpected happenings and reformations. [2, pp. 51–52]

Since [2] is a book on numerology and not on number mysticism, the numbers have *applications*:

> Take another case. You have to call upon a firm and you hope to pull off a good piece of business while there. You are booked to reach the firm by 2.30. Thoughts of numerology flash across your mind and, on the back of an old envelope, you scribble the now familiar inverted pyramid, but with these figures:

```
2     3     0
   5     3
      8
```

> "Ah!" you say, "that's no good, as my number is *two* and not *eight*." Well, if your number is *two*, why not reach there at 2.25, and all will be well, as the following pyramid shows

```
2     2     5
   4     7
      2
```

> Mind you, we are not suggesting if you were going to Edinburgh, and a train at 2.30 suited your convenience, but did not give your number, and that an inconvenient train at midnight gave your number, that you should lose the first and catch the later one. Numerology, when used in this way, is more a nuisance than a help; and its real mission is to help.
>
> Do not think that numerology is something to use now and again. It can be applied to the drawing-up of a dinner menu, a dance programme, to mapping out a summer holiday or buying a new frock. It uses are unlimited.
>
> [2, pp. 64–65]

Of course, numerology must be used with care. If you arrived for your 2:30 appointment at 2:41, the numbers would be right

```
2   4   1
  6   5
    2
```

but you might lose the business on account of being late.

Number-triangles did not catch on, and now everyone's number is determined by straightforward addition of the digits in the name. The characteristics of numbers are also now standard. In Table 1, from a numerology book chosen essentially at random [1, pp. 25–29], are the key words that go with each digit.

Note the similarities to the characteristics given by Iamblichus. Seven is the number without parents; Athena, who sprang from the forehead of Zeus, had no parents; Athena was the goddess of wisdom; hence seven was, and is, the number of wisdom. Thus does the logic of number mysticism operate.

1	INDIVIDUALITY (Will)
2	COOPERATION (Peace)
3	SELF-EXPRESSION (Creativity)
4	DISCIPLINE (Work)
5	FREEDOM (Change)
6	RESPONSIBILITY (Service)
7	WISDOM (Detachment)
8	AUTHORITY (Power)
9	ALTRUISM (Compassion)

TABLE 1 Numerological characteristics of digits.

Actually, it is not logic but magic. One of the fundamental principles of magic is that like affects like. Stick a pin in a doll and you hurt the person in whose image the doll is made: like affects like. Athena is like 7, thus Athena affects 7. But this is not the place to go into the postulates and theorems of magic.

Note also the similarities of the characteristics of the digits to those of the believers in the enneagram (chapter 28), who avoid mentioning the occult in their literature. Thus does number mysticism spread.

In case you do not yet have a firm grasp on the personalities of the integers, here they are, from [2, pp. 25–29], arranged by yang, yin, and their combination. When you have mastered them, you will have mastered most of the content of most of the numerology books on the market today. That is no exaggeration. Numerology books consist of a vast amount of elaboration, with very little content, of the mystical properties of the digits.

1

Assertive—willful, domineering, selfish, arrogant: puts own needs before others regardless of personal consequences; boastful, impulsive.

Passive—dependency, submissive, fearful of making decisions or taking initiative, stubborn, procrastinating.

Harmonious—strong willed and ambitious with consideration for others, courageous, organizer-leader-pioneer, individual, original thinker.

2

Assertive—meddling, arbitrary, careless, strident, tactless, extremist, dishonest; overlooks detail, creates divisiveness.

Passive—vacillating, sullen, devious, faultfinding; too much attention to detail with delay in accomplishment; inability to take a stand.

Harmonious—diplomatic, adaptable, able to fuse divergent opinions or groups, rhythmic, gentle; gathers information from both sides of a position before taking a stand.

3

Assertive—superficial, extravagant; likes to gossip, has false vanity, wasteful attachment to ego gifts, gaudy taste; dislikes the practical.

Passive—lacks concentration; failure to fully develop creative potential; asexual, gloomy, lacks imagination.

Harmonious—gift of speech, creative and artistic, intuitive, joyful, sociable, enthusiastic, tasteful in dress and decorum.

4

Assertive—stubborn, sees one way of doing things, intolerant, too serious, brusque, lacks emotional sensitivity, overworks.

Passive—lazy, resists new methods, narrow, must see practically before accepting ideas, fights intellectualism.

Harmonious—loyal, consistent, patient; fulfills a given task well, sticks to facts; organized, economical, has integrity.

5

Assertive—restless, disregards values, irresponsible, nervous, overindulgent (particularly with senses); has too many interests, ignores rules and laws.

Passive—fears the new or change, wants a rule for every behavior; sexual confusion, failure to learn from experience, uncertain.

Harmonious—progressive, has diverse talents and friends, is curious, seeks freedom; quick, flexible, adventurous, energetic; enjoys travel.

6

Assertive—over-involvement with others' problems, self-righteousness; is worrisome, domestically dictatorial, prone to arguing, overly conventional, easily upset.

Passive—martyrs himself, resents service, lacks concern for family and home; constant complainer, anxious, carries too many burdens (real or not) on shoulders.

Harmonious—has harmonious home, serves humanity, is unselfish, artistic in taste, conscientious, fair; seeks emotional equilibrium in self and others.

7

Assertive—severely critical, too analytical, has intellectual conceit and vanity; deceptive, aloof, eccentric, faultfinding

Passive—skeptical, has inferiority complex, cynical, suppressive, cold; thinks rather than acts; crafty, prone to emotional withdrawal, secretive.

Harmonious—excellent analyst, seeks deeper truths, has technical ability, reaches, has faith; mystical and intuitive if higher mind is tapped; has stoic temperament, is discerning, has poise.

8

Assertive—over-ambitious in order to attain leadership or power; has callous disregard for others, is crassly materialistic, demands recognition, loves display; abusive.

Passive—fears failure, unable to take a leader's role, has poor judgment; scheming, disrespects authority, dishonest in business, has careless regard for money.

Harmonious—has executive ability, respects wealth; successful, good judge of character, confident, administrates with personal consideration for others.

9

Assertive—impractical, fickle, over-idealistic, lacks tolerance for others; views, too generous, indiscreet, anarchists at extremes.

Passive—aimless, gullible, easily used by others, depressed, indifferent, pessimistic about the world and the future.

Harmonious—inspired, attuned to New Age concepts, compassionate, generous, gifted artistically; seeks world brotherhood and harmony, is a perfectionist, works to build group consciousness.

There, you are now ready to set up shop as a numerologist. One way to do so is given in the next chapter.

References

1. Buess, Lynn, *Numerology for the New Age*, Light Technology, Sedona, Arizona, 1978.
2. Numero, *The Power of Numbers*, David McKay Co., Philadelphia, no date.

CHAPTER 21

What Numerologists Sell

Let us look at a numerological analysis, and try to see why it is that people will continue to pay for them and continue to think that there is something to them. The one we will consider was computer-generated, making it all the more impressive.

Computers have been pressed into the service of numerology. There are several numerology programs on the market, at least one aimed at the professional numerologist. Here is an advertisement for it:

> [This program] is designed to provide quick and easy analysis without the need to flip through numerous books and charts.
>
> $$ THE PROFESSIONAL VERSION $$
>
> Yes! You can make hundreds, even thousands, of dollars at shows, fetes, weekend markets and shopping malls by providing attractive personal analysis outputs on pre-printed forms. For only $90 US ($120 Canadian) you can get started! This package includes the money-making program together with full documentation, list of equipment required and simple pre-printed form. Programs and instructions are given to allow you to customize the program output and configure the program for use with different printers. The program even keeps count of the number of printouts made and provides impressive computer graphics to assist in attracting customers! Other features of this program include; day of birth information, a lucky number printout and an options field which allows you to override certain print fields on a customer to customer basis.

The analysis that follows was not taken from that program, but from another, *By the Numbers*. My copy of it is version 4.41, the version number showing that it has had long, wide, and successful use. Let us start by looking at each sentence in the analysis and estimating its content. By "content" we will mean, roughly, the probability that a person chosen at random would say, "That doesn't apply to me." A sentence that makes a statement that no one could conceivably disagree with has content 0. The sentence "You are Carl Friedrich Gauss." has content as close to 1 as makes no difference.

The program first prints out the numerological significance of your last name.

> THE IMPORTANCE OF YOUR FAMILY NAME: 8
> *The number 8 family emphasizes hard work and practicality.*

It is conceivable that the person reading this statement could disagree with it, but the chance is small. There are families that are impractical and some in which children are told not to work hard, but I think that among the clients of numerologists their proportion is not large. It is generous to give this sentence a content of .1.

> *If the atmosphere is negative, the parents will tend to be domineering and materialistic, and to denigrate a child's playful nature.*

Given the preceding sentence, this has content 0.

> *When the atmosphere is positive, the parents set an example of good judgement and money management, and the family learns to work together as a team.*

Similarly, no content. Conditional statements always have less content than assertions.

> *There may be a medical history of headaches, arthritis, and age-related problems.*

The "may" drains a good deal of the content out of this sentence. Everyone has headaches, some luckily not often, and anyone who lives to even the borders of old age is going to have age-related problems. Arthritis, though common, is not universal, so this sentence has some content, though not much. Shall we say, because of the "may," .05?

> *Your leadership skills and organizational mind are useful in almost any undertaking.*

Well, of course. Do people admit, publicly or privately, to a complete lack of leadership skills, or to the possession of a disorganized mind? Not after a numerologist has told them that they have those admirable qualities, they don't. Zero content.

> *You may find yourself particularly attracted to a career in business or finance.*

Notice the cleverness of numerologists. It is probably instinctive, but it is cleverness nevertheless. It would not do to say that you have, or will have, a career in business. That statement has positive content, though the proportion of people who have careers in what could be described as "business" is large. Anyone who works for an organization whose purpose is to make a profit is in "business." Nor does the sentence say that you *are* attracted to a career in business or finance. There are people, even those whose family number is 8, who are not so attracted. So, the numerologist says that you "may" feel such an attraction. That gets the content down. 0.1 is probably generous.

> *Whatever career you choose, you will approach it with determination.*

I should think so! Some sort of determination, certainly. If not the determination to succeed, then the determination to make a little money, or the determination to avoid the disagreeable parts of the job. Determination, being vague, is an easy quality to give yourself.

> *Don't let your seriousness detract from, your normally friendly nature.*

The content of this sentence is determined by the proportion of people who think of themselves as either frivolous, hostile, frivolously hostile, or hostilely frivolous. Though there are such people, the proportion who think of themselves that way is very small. For that matter, if you have no seriousness at all, then it cannot possibly detract from anything. Since I think that some people do not think of themselves as serious this sentence has a non-zero content, but no more than .03.

> *People will look to you for authority as long as you do not go overboard and become domineering or greedy.*

Another .03, since there are some people so beaten down by life that they do not think of themselves as wielding any authority at all. However, they probably do not consult numerologists.

> *Your mature attitude makes it easy for you to get along with those older or more experienced than you, and allows you to be patient and persuasive to those who are younger.*

Flattering words like these flow over us like honey and, like honey, we eat them up. It is too bad that they have no content.

If my estimates of content are correct, then the probability that *any* reader of that analysis would agree with all of its statements is .80, rather high. Even if one statement missed the mark by a bit, the average consumer of numerology would disregard it because of the amazing accuracy of the rest.

There follows the remainder of the analysis. The estimation of its content is left as an exercise for the reader.

THE INFLUENCE OF YOUR FIRST NAME

> Your first name is probably the strongest influence in your daily life. Not only does it describe your outward personality and the way others see you, but your own response to it reflects your innermost desires and impulses.
>
> Your name adds up to number 2. The 2 is characterized by a peaceful and cooperative nature. Its strengths are in diplomacy, culture and charm. Its weaknesses are indecisiveness, shyness and passivity. You will probably be attracted to a career that allows you to work in harmony with others, perhaps as a liaison between groups or in a position of offering hospitality. Your easygoing nature balances a tendency to be hypersensitive or petty about details. People are drawn to your quiet charm and modest appearance.

Note the proper Neopythagorean properties of 2. Note also that they are applied to people, something the Neopythagoreans never did. This is the advance that Mrs. Balliett made past Iamblichus and his fellow Neopythagoreans.

OTHER IMPORTANT FACTORS

> Because your most significant number, the number of your first name, is the same as your personality number, many of these characteristics will be emphasized accordingly.
>
> In matters of love and marriage, it is important to remember that extreme opposites do not attract. At the same time, too much similarity can stifle growth in the relationship and can also limit you as an individual. Basic similarities and complementary differences are the key to balance.
>
> Similar birth numbers and destiny numbers are considered important signs of compatibility. Additionally, you will find yourself attracted to people whose birth dates add up to 2, 3, 6 or 9. You will tend to have problems with those whose birth dates add up to 4 or 8.

Birth number, destiny number, personality number, first-name number, last-name number: when you have a lot of numbers, you will have a lot of Neopythagorean properties ascribed to you and many of them will apply to you. You will tend not to notice the ones that do not, especially when they are presented in numerological fashion: "you will *probably*. . . ," "*perhaps* as a liaison," and so on.

> Many people find you attractive and charming. Although you do not fall in love often, when you do, you are quite devoted. You feel most romantic just before and just after the weekend.

YOUR PLANES OF TEMPERAMENT AND POWER

> Your name is highest on the PHYSICAL plane. Things that are tangible, physically real, have more meaning for you than things which seem metaphysical or imaginative. You have good physical endurance as well as mental stamina. Your nature is practical, probably leaning toward the conservative.

You have a good sense of responsibility, but you may become restless if your work seems too repetitive. You enjoy a change of pace now and then.

OTHER COMPONENTS OF YOUR CHARACTER

Your EXPRESSION number, 1, describes the combination of your first and last names, and thus gives an overall picture of the total you. This number will dominate over those of your PERSONALITY, 2, and INSTINCTIVE DESIRE, 9.

EXPRESSION: The number one is very forceful. Its positive characteristics are originality, ambition, independence, and daring. Its negative traits are selfishness, domination, and impatience.

PERSONALITY: Two is a gentle number. Its best traits are diplomacy, charm, sensitivity, and cooperation. Two's negative traits are shyness, indecisiveness and moodiness.

DESIRE: The number eight's positive traits are leadership, financial ability, and determination. Its negative qualities are greed, hate and oppression.

YOUR DESTINY AND FORECAST NUMBERS

Your destiny and forecast numbers describe the lessons you need to learn in life, and offer a view of situations that lie ahead. It is important to remember that a forecast is not a hard and fast prediction of events. The forecast indicates the most probable direction for you and the type of circumstances you are most likely to face, given your overall character and traits.

Your ability to learn from experience, to anticipate situations and feelings, will allow you to make the most of the opportunities that come your way, and give you the chance to avoid or solve problems. With Numerology, you are not a helpless pawn of fate. Rather, your increased understanding and awareness allow you to take charge of your own destiny.

YOUR PERSONAL FORECAST FOR 1994

1994 is a 3 year for you. Imagination and creativity are emphasized. This appears to be a very happy year in your life, with much social activity and leisure. It will be important for you to concentrate your energies in order to fulfill your creative potential. Don't be too distracted by all the fun that comes your way. You can be popular and have a good time and still complete those important projects. Any sort of artistic endeavor or writing is especially favored. You may find yourself engaging in some flirtations, but they are not likely to be serious. Be selective in your friendships.

Your cornerstone is 3, the same as your forecast number for 1994. This can give you some extra inspiration and a special boost to your youthful enthusiasm.

YOUR LUCKY NUMBERS: 6 2 9 3

Your lucky colors are BLUE and COPPER.

That was a short analysis, chosen to spare the reader. Too much unrelieved vapidity can cause dangerous pressure to build up, which may be released in unsocial acts such as screaming or throwing things. If you doubt that it is vapid, ask yourself if it does not apply to you. You cannot disagree with "Be selective in your friendships," can you? The entire analysis has content close to zero.

Another program, designed to give customers their money's worth, printed out eight single-spaced pages after I gave it the name of Carl Friedrich Gauss and his date of birth, April 30, 1777. It included such insights as

> There's a charming and gracious side of you which is most likely to be displayed when you're comfortable with close friends. At times when you feel particularly uncomfortable, though, you can be critical, oversensitive or moody. You have a good imagination and may enjoy creative pursuits. You probably possess good verbal ability and it wouldn't be surprising if you sometime involve yourself in activities related to words. You may have some singing or acting talent, or you may enjoy writing or lecturing. You're probably a good conversationalist when you care to show that side of yourself.

Gauss breaking into song! What a picture! But it is possible that he could think, even as could you or I, "yes, that's right, and no one knows about that except me." Even so, Gauss, being Gauss, would not therefore conclude, and neither would you (I hope), that numbers know our secrets and determine our character. However, enough people do so conclude, or half-conclude, to provide a steady flow of customers to sellers of numerology. I wonder if Pythagoras would have approved.

If you would like to have your own numerological analysis, opportunities abound. Many entrepreneurs have numerology programs and would be happy to send you what they have to say about you. For a price, of course. It will cost you $24.95 to get

> 30 to 35 computer printed pages [which] include your Heart's Desire, Life Path, Hidden Passion, Balance Numbers, Destiny, and much more.

However, three analyses come for only $49.90. One numerologist in California advertises

> If you would like to have your Individual or Compatibility reading done please send $38.00 certified check or money order to [the numerologist]. Please allow 2 weeks for completion.

A personal check is not good enough!

The following sounds more appealing.

> NUMEROLOGY describes the journey of the soul, and as you grow in consciousness along the years you yearn for a deeper-more rewarding life.... You will be able to reach a higher inner attainment, to love yourself and other like never before—and, finally, experience this blissful Higher FREEDOM.

But bliss comes high—your Personality Chart will set you back $65. Perhaps a better bargain, for $23.50 plus $1.50 postage and handling charges, is a report that includes

- ☐ Life Path—The major lesson you are here to learn.
- ☐ Expression—Your special inborn abilities and talents.
- ☐ Soul Urge—Your inner motive.
- ☐ Birthday—A minor lesson in addition to the Life Path.
- ☐ Master Numbers—Special energies indicating spiritual opportunities.
- ☐ Repeated Numbers—Energies that may be out of balance.
- ☐ Karmic debts—Areas of difficulty because of the misuse of an energy in a past life.
- ☐ Karmic lessons—Weaknesses from lack of experience in previous lives.
- ☐ Intensity Points—Strengths or weaknesses encountered because of too little or too much of a particular energy.
- ☐ Challenge—An obstacle encountered early in life.
- ☐ Maturity Number—An important lesson learned later in life.
- ☐ Current Name Energies—New or increased influences from a name change.

People must buy this stuff. Presumably people even send money to the numerologist who advertises:

> Unlike cheap versions of give me your first and your last name and receive your "personalized" reading in a minute we based my entire "calculation engine" on three numbers which make you truly unique from anyone else in the world. We use your birthday as well as your parents birthdays to give you your unique reading for any given month in any given year. Our system is the only system in the world which follows Pythagoras instructions leter by leter and doesn't do any averaging of your numbers!

Yes, "leter by leter." Literacy is no prerequisite for numerology. Numerology is all about numbers, after all.

CHAPTER 22

Listen for Your Number

Can anyone explain why some movements are brittle and others are not? That is, why some are prone to schisms and sects and others hold together? The United States seems to have as many varieties of Baptists as Heinz has of pickles, but Mormons and Christian Scientists have held together in one piece. Since one of those two churches is flourishing and the other is dying, the health of an organization is no guide to its frangibility. Some woods split easily and others do not, and so it is with groups. Which way they go may well depend on nothing other than the blind workings of chance.

On a scale of 1 to 10, with 10 being maximally monolithic and 1 being shatterable at a glance, numerology is right up there around 9 or so. Mrs. Balliett did her numerological work so well that future laborers in the field have not found it necessary or desirable to depart from her teachings.

There is one exception, the only one I know of, whose existence keeps numerology from scoring a perfect 10. A tiny book (6 3/4″ by 4 1/4″, 56 pages) by Mabel Ahmad, published in 1924, is dedicated to the author's late husband, S. H. Ahmad:

> In 1903 he expounded and proved the operation of the practical Law of Numbers, et cetera, to the Western World, for the first time in history.
>
> [1, p. v]

Ms. A.'s book is not an exception to the numerological axiom that a person's name has enormous influence over character, personality, and so on. That is too basic to be tampered with. Naming, for numerologists and other groups of primitive people, is mystically equivalent to creation, and the names of people are more important even than the astrological sign they are born under. Nor is it an exception to axiom two of numerology theory, which is that names can be reduced to numbers. As we have seen, the Balliett system, the one used without exception by modern numerologists, looks at the letters of the name, assigns them digits, sums, and reduces the total modulo 9, replacing

1	2	3	4	5	6	7	8	9	10	11	12
A	B	C	D	E	F	G	H	I	J	K	L
M	N	O	P	Q	R	S	T	U	V	W	X
Y	Z										

TABLE 1 New numerological letter values.

0 with 9. Some numerologists refuse to reduce 11 and 22 to 2 and 4, but that small departure from orthodoxy is not enough to count as a schism.

By the way, it is strange that no numerologist has ever had the idea of assigning the letters of the alphabet values based on 12, as shown in Table 1, and then reducing the sum of values of the letters modulo 12 instead of modulo 9. So, for example, George Washington would have 4 for his number. From Table 1, George has value $7 + 5 + 3 + 6 + 7 + 5 = 34 \equiv 1 \pmod{11}$ and Washington is $11 + 1 + 7 + 8 + 9 + 2 + 7 + 8 + 3 + 2 = 58 \equiv 3 \pmod{11}$. Thus George Washington is $1 + 3 \equiv 4 \pmod{11}$.

The huge advantage of this system would be that numerologists could then make connections between name numbers and signs of the zodiac. Think of the implications of having your number the same as the number of your astrological sign! Or of having it six numbers away! I feel mystical insight coming to me. Clearly, to have the same number as the number of your sign would be good, since all is vibrating in harmony. Having a number six numbers away from your sign would also be good, though in a different way, as thesis and antithesis produce synthesis. Having a number *three* numbers away would not be so good because of the discord and pulling at cross purposes. Being one number away would be even worse, being as discordant as playing a C and D simultaneously on a piano.

That may not be entirely coherent, but then coherence and mysticism have very little to do with each other. However, the idea is a natural and a book based on base-11 numbering would write itself. Even devoting only two pages for each name-number–astrological-sign combination (and any astrolonumerologist could easily go on for two pages), you have 288 pages already. Besides numbers, there are the characteristics of the astrological signs to consider, with the implications of being a fiery Sagittarius whose number is the watery 2 to be worked out. Add an introduction, an explanation of how to find your number, the characteristics of the twelve name-numbers, and a few remarks and you have a hefty and impressive 350-page book, ready to sell thousands and thousands of copies. And it would. But only if you really believed what you were writing.

The reason that no one has seized on this idea yet may be that numerologists would find it too hard to reduce numbers modulo 11. Or it may be that they have no new ideas and all copy from Mrs. Balliett. But the best reason is that it is not true to Pythagorean number mysticism, where it is the digits that have properties and not the numbers from 1 to 11.

It is ancient Pythagoreanism that was at the base of Ms. A.'s revision of the Balliett system of assigning numbers to names. Ms. A. (or her husband) remembered Pythagoras's music of the spheres. Numbers make noise! Since all is number, numbers vibrate, and sound is vibration, it follows that numbers are sound and maybe even that all is sound. Ms. A. does not go quite that far:

> The forces represented in the vibrations caused by the known perpetual revolutions in this solar system are intelligible to man through the sounds produced. These are susceptible to numerical calculation, although of quite different significance to that of arithmetic. The wheel of a motor, according to its speed, vibrates, as even a child knows, to a sound. A can of water swung swiftly in a circle, above and under the arm, not only produces a sound but retains, at sufficient speed, the water—hence we have sound and order. So it is with the universe. Certain bodies have a heavy, slow ponderance, and produce sound accordingly, and others are lighter and quicker in revolution, producing lighter sounds.
>
> These vibrations and sounds have their counterparts in varying degrees in every man, beast, and place; these are recorded in the name (which is not received by chance). [1, p. ix–x]

Ms. A. had the idea that it is the *sound* of the name that is the key to its numerical value, and not its letters. Of course, number is what is important:

> The fact that the revolutions of the main planets are numbered—that every natural relation in life is susceptible to number—the scientific arrangement of the seven days in the week, the natural twenty-four hours of the day—sixty minutes to the hour—the four seasons, the lapse of nine months before birth, and so on, may open the eyes to a glimmering of an idea that such a Law is in existence. [1, p. vii–viii]
>
> Higher science has become subject to the wonderful advances made in mechanical science, to which the Law of Numbers is really the key. Numbers are the symbols of the forces through the medium of which we are governed by the Great Architect of this universe.... Therefore it matters to the individual that his home, partner, friends, etc., should have names harmonizing with his own. The letters in a name are only the symbols of the ideas represented in the name, and by the alphabetical explanations within these covers any man can deduce the numerical character of his or any name. [1, pp. ix–x]

This is not entirely clear, but clarity is not to be expected of a writer who thinks ("the natural twenty-four hours of the day") that the numbers on a clock are as inevitable as the nine months of human gestation. But, she is clear on the system:

> Remember, it is the sounds that count, not the number of letters; Nature is never the subject of the whims of mankind. LONDON is pronounced LNDN and must be enumerated thus. Double M is always pronounced as one M—double L as one L—and COUGH is by sound COF.
>
> [1, p. xiv]

Also clear is that Mrs. Balliett was wrong:

> Many interesting strange books, by different writers, purporting to number names, have appeared, but I must warn the reader that they have no connection with this effort, which is based purely on the Law and not on imagination.
>
> [1, p. xiii]

Naturally, no reasons are given for the operation of the Law. Ms. A. says to try it out and you will be convinced:

> No one will be skeptic enough, when he has tested some names for himself, to doubt that the successful harmonies in the numbers of the names and residences given above are according to Law and not the result of Chance. We are part of a whole, and the forces with the greatest influence in play on everything in the life are registered in the name. [1, p. xi–xii]

The Ahmad system is based on the 22-letter Hebrew alphabet, which is more satisfactory—it is more ancient, more impressive, more mystical—than the English 26-letter alphabet, for Pythagoras knew nothing of modern English. He knew nothing of ancient Hebrew writing either, but it is at least ancient. Ms. A.'s phonetic system is given in Table 2.

Now comes the difficult task of translating English sounds to their closest Hebrew equivalent so as to be able to give them numerical values. For example, here is what Ms. A. makes of *a* sounds. Given *a* as in *father*, *man*, *lark*, *addition* ("as an open sound and as the initial sound where slighted"), assign the number 1. Broad *a*, however—*lane*, *Mary*, *relate*—is 10, as is the *a* sound in *hair*, *tear*, and *phæton*. As an emphasized initial sound, as in *Amy*, *aquatic*, and *alien*, *a* is 11. If doubled, as in *Baalbek*, it is 2. Further, there are other letters that can have value 1, as a very short initial *u* (*uncle*, *umbrella*) or clipped *i* (*inky*).

Consonants are easier, but even there we have to be careful to distinguish the *th* in *thing* (9) from that in *lathe* (4) and the *tt* in *Betty* (9) from that in *Brett* (400). The vowel in *bowl* has value 6, in *cow* 7; the *u* in *tune* is 16, in *run* is 0, and in *prune* is 6.

This complexity goes a long way to explaining why the Ahmad system of numerology never caught on. More than half of Ms. A.'s book is devoted to explaining the correspondences of sounds with numbers. An Ahmad-numerologist could not give a client a number with anything like the rapidity of a Balliett-numerologist. Clients might even be put off if their numerologist

Serial Number	Phonetic Value	Numerical Value	Name
1	A	1	Aleph
2	B	2	Beth
3	G	3	Gimel
4	D	4	Daleth
5	H	5	He
6	W and V	6	Vau
7	Z	7	Zayin
8	h	8	Cheth
9	Th	9	Teth
10	Y	10	Yod
11	K	20	Kaph
12	L	30	Lamed
13	M	40	Mem
14	N	50	Nun
15	S	60	Samech
16	'A (Arabic)	70	'Ayin
17	P	80	Pe
18	Ts	90	Tsade
19	Q (Arabic)	100	Qoph
20	R	200	Resh
21	Sh	300	Shin
22	T	400	Tau

TABLE 2 Sounds to numbers.

had to refer to a book to get the analysis under way, though with study and practice the book could be dispensed with.

The Ahmad system raises the question of what is to be done about the mutability of pronunciation. Spelling also varies, but not as much as the sounds of words in the mouths of different speakers. Does one's character change with one's accent? Ms. A. says, "Nature is never the subject of the whims of mankind," but pronunciation can be whimsical. Most people I know say the *th* in *thing* and *lathe* the same way, the only difference being that the second *th* is voiced and the first is not, but that is not the reason why the first *th* has value 9 and the second 4:

> TH when tending to a hard sound like that of D takes a weight of sound values, as we have seen, of 4. [1, p. 16]

Ms. A. must have pronounced *lathe* as *lade*, or maybe *ladthe*. Also, she must have said *Betty* something like *Bethy*, since the value of 9 for that *t*-sound came because the "TT in the centre of a word has the sound value of TH." [1, p. 17]

In spite of how pronunciation changes with time, Ms. A. has a valid numerological point. If a person *is*, in a mystical sense, a name, and the name, in a mystical sense, vibrates to a number, then the sound of the name seems to be more fundamental than its orthography. The symbols used to represent sounds are arbitrary, and vary. Do people change their characters when they move to a country where spelling is different? Maybe they do, and maybe they don't. Maybe it all depends on how people *think* of their names. And maybe personalities *do* change with pronunciation: traditionally, *Smiths* and *Smythes* are different sorts of people and act in different ways. Of course there the spellings are different as well. Perhaps each of us has a Platonic ideal pronunciation of our name that we can only approximate. It is all very difficult. More research is needed.

Besides being difficult to apply, another reason the Ahmad it's-the-sound-that-counts doctrine has not gained adherents may be that Ms. A. gave so few examples of its application. Pointing out that *Wembley*, *empire*, and *London* all have value 8 (from 98, 332, 134, respectively, [1, p. 46]) or that *tap*, *water*, *drink*, and *glass* are all 4 (481, 607, 274, and 94, [1, p. 51]) does not do much for those who want to know what their numbers tell about their characters.

For whatever reason, the Ahmad system is now dead and numerology presents a united face to the world. Numerology has no heretics.

Reference

1. Ahmad, Mabel L., *Names and Their Numbers*, David McKay Co., Philadelphia, 1924.

CHAPTER 23

The Power of the Pyramid

If you have ever seen the pyramids of Egypt, you have been impressed. They are *big*. They are also old. They have been standing there for four thousand years and they look good for several millennia yet. In their presence, you might have a semimystical experience by meditating on their size and age, though the importunations of the sellers of postcards and camel rides make meditation difficult.

Mystical experience or not, the pyramids, especially the largest of them, the great pyramid, have affected the minds of many people. They have looked at the great pyramid, or read about it, and decided that such an immense and impressive structure must have a *purpose*, also immense and impressive. The great pyramid, they think, is not what it seems. They know that it contains secrets. They become pyramidologists.

Pyramidologists assert that the great pyramid was put in a very special place, for very special purposes. Some find the history of the world in its measurements (see the next chapter), while others, who do not go that far, find φ ($= 1 + \sqrt{5})/2$, the golden section, see chapter 29) and π in it, and many other wonders. They present volumes of evidence. They are convinced, and they convince others that they are right. Pyramidology lives.

Could it nevertheless be that the pyramid had no purpose other than to be a monument and was built with never a thought for φ or π? Since the ancient Egyptians knew nothing about φ and hardly anything about π, this seems very likely to me.

It seemed likely as well to Willy Ley (1903–69), a science writer and rocket enthusiast, a native of Germany who wrote English considerably better than most native-born writers of the language. In "The great pyramid, the golden section and pi," he explained the construction of the pyramid to my satisfaction, and showed how it led the pyramidologists—misled them, actually—to finding φ and π in it.

Here is what he had to say about its location.

> Why Khufu picked this particular locality for his pyramid is not known. That is, we don't know of any inscription saying that Pharaoh decreed this site because his fellow gods had told him to begin his soul-voyage after death from this place. But we can think of a number of eminently practical reasons why he chose this spot.
>
> To begin with, he could see it from his summer palace and watch the actual work going on. Secondly, the location of the building site was such that, when the annual Nile flood occurred, the blocks of stone could be floated on rafts to the foot of the growing structure. Finally, most egyptologists (and especially Egyptian egyptologists) believe that Khufu's pyramid hid an outcropping of natural rock which obviously saved that much work in the erection of a virtually solid structure. [4, pp. 27–28]

That last point is not only eminently practical, it also can explain why the great pyramid is the largest pyramid ever built: Khufu grabbed the outcrop and there was none left for later pyramid-builders. The two later pyramids that share the Giza plateau with the great pyramid, though smaller, may have taken *more* work to construct.

As for the motives of King Khufu (or Cheops, to give another common rendering of his name), we can only speculate. But speculation provides reasons for constructing the pyramid that are convincing enough that we do not have to appeal to space aliens, supernatural beings, or possessors of lost wisdom to explain its existence. Earlier kings had built pyramids, and had been building them for centuries. Pyramids were an ancient Egyptian art form; how natural for a king to want to carry it a step further and create the largest and most impressive structure on earth, one which would survive for millennia. What an excellent way to guarantee immortality! And it has worked. However his name is spelled, Cheops or Khufu, it will be known long after our names are forgotten.

There was in addition a political reason for building the pyramid. Keep a population occupied, or exhausted, and it will have neither the time nor the energy for revolution. Cheops was evidently not the most popular of rulers, as Herodotus (c. 480 B.C.) tells us [2]:

> Down to the time when Rhampsinitos was king, they told me there was in Egypt nothing but orderly rule, and Egypt prospered greatly; but after him, Cheops became king over them and brought them to every kind of evil; for he shut up all the temples, and having first kept them from sacrifices there, he bade all the Egyptians work for him.

And work they did:

> So some were appointed to draw stones from the stone-quarries in the Arabian mountains to the Nile, and others he ordered to receive the stones after they

> had been carried over the river in boats, and to draw them to those which are called the Libyan mountains; and they worked by a hundred thousand men at a time, for each three months continually. Of this oppression there passed ten years while the causeway was made by which they drew the stones, which causeway they built, and in a work not much less, as it appears to me, than the pyramid; for the length of it is five furlongs and the breadth ten fathoms and the height, where it is highest, eight fathoms, and it is made of stone smoothed and with figures carved on it.

After ten years, work could begin:

> For the making of the pyramid itself there passed a period of twenty years; and the pyramid is square, each side measuring eight hundred feet, and the height of it is the same. It is built of stone smoothed and fitted together in the most perfect manner, not one of the stones being less than thirty feet in length. This pyramid was made after the manner of steps, which some called "rows" and others "bases": and when they had first made it thus, they raised the remaining stones with machines made of short pieces of timber, raising them first from the ground to the first stage of the steps, and when the stone got up to this it was placed on another machine standing on the first stage, and so from this it was drawn to the second upon another machine; for as many as were the courses of the steps, so many machines there were also, or perhaps they transferred one and the same machine, made so easily as to be carried, to each stage successively, in order that they might take up the stones; for let it be told in both ways, according as it is reported.

That sounds straightforward and reasonable. Notice that Herodotus's informants do not mention mysteries, magic, π, or φ. They report only that the pyramid was built. This is significant for, then as now, the temptation of the natives to tell a story to see how the inquisitive but ignorant foreigner will take it was not to be resisted. Herodotus passed along the following whopper:

> Cheops moreover came, they said, to such a pitch of wickedness, that being in want of money he caused his own daughter to sit in the stews, and ordered her to obtain from those who came a certain amount of money (how much it was they did not tell me); and she not only obtained the sum appointed by her father, but also she formed a design for herself privately to leave behind her a memorial, and she requested each man who came in to her to give her one stone upon her building; and of these stones, they told me, the pyramid was built which stands in front of the great pyramid in the middle of the three, each side being one hundred and fifty feet in length.

If there had been mysterious signs and wonders at the pyramid one would think that there would be better stories to tell than that fairly feeble attempt to be amazing.

It was not until the nineteenth century that the amazements began to appear. Modern pyramidiocy (though that is not what the pyramidologists call it) started in 1859, with John Taylor's *The Great Pyramid, Why Was it Built and Who Built it?* [8] Taylor rejected the obvious answer that it was built by an ancient Egyptian king for the purposes of preventing rebellion and leaving an enormous monument to himself. Ley says that

> [Taylor's] conclusion was that the pyramid had been built for the purpose of embodying a few important measurements. If Taylor had been an American, he might have said that it was the Egyptian equivalent of the Bureau of Standards, with the additional twist that all the standards are "classified information" not meant for the average dumb citizen....
>
> [He] found that the height of the pyramid (which he overestimated by a few feet) was $1/270,000$ of the circumference of the Earth. Here one can only say, "Why not?" for the height of the pyramid must be some fraction of the circumference (or of the diameter) of the Earth. [2, p. 31]

When pyramidologists find astonishing numbers in the pyramid, it should always be kept in mind that they have found them by going the wrong way around. If Taylor had said, before looking at any measurements, "I hypothesize [for whatever reasons] that the height of the pyramid will be an even fraction of the earth's circumference" and *then* found his 1/270,000, that would be astonishing and worth paying attention to. But he had the measurements first. See chapter 10 for an example of a similar maneuver—pulling 57 out of a dollar bill. Taylor had the choice of the equatorial circumference, the polar circumference (which is different since our planet bulges in the middle), or the circumference of any great circle—one passing through the base of the pyramid would seem to be a natural choice—and the choice of the pyramid height, length of base, slant height, altitude: it would not be hard to find two numbers whose ratio was a round number like 270,000 to 1. Since he could always alter the numbers by a few feet here and there, not only would it not be hard to find a round-number ratio, it would be hard *not* to.

> Taylor was willing to admit that a few such things could be coincidences. To feel sure that it was planned, he looked for something which could not be a coincidence—or so he thought—and he found one, too. The square of the height of the pyramid, compared to the area of one of its triangular faces, demonstrated the "golden section."

The fact that the Egyptians knew nothing at all about φ did not bother Taylor, any more than their ignorance about the circumference of the earth. The number that he calculated seemed to be close to 1.618, though accuracy to four significant figures was not justified by the accuracy of the numbers that produced them, so he concluded that it was *exactly* φ ($1.6180339887498948\ldots$). After all, everyone knows that things with φ in them are beautiful.

By the way, that is another of the many notions about φ that are (a) believed by everyone and (b) false. Everyone knows (see chapter 29) that the golden rectangle—the rectangle whose sides have the ratio φ to 1—is the most esthetically pleasing rectangle. If golden rectangles are so beautiful, why is it that books, surely designed to be as pretty as possible with the hope that they will jump into a purchaser's hand (no one designs a book to be ugly) do not display it? Taking a row of nine books off my shelves at random (nine because there are nine digits—as number mystics know, there must be a reason for every number), I find the ratios of their heights to their widths to be, in increasing order,

$$1.251, 1.456, 1.470, 1.507, 1.515, 1.528, 1.535, 1.539, \text{ and } 1.631.$$

Not one of the ratios is close to φ, nor is their mean. The crudest of statistical tests allows the immediate rejection, at the 5% level, of the hypothesis that the books were taken from a population whose mean ratio was φ. The book with the height-to-width ratio of 1.631 looks ugly. It is too tall, and too thin. No wonder it was remaindered.

The discovery of φ did not take place until long after the capstone was placed on the last pyramid. A theory of proportion and geometry was needed, things that the ancient Egyptians did not have. It had to wait until the ancient Greeks invented them. Greek civilization gave way to that of the nonmathematical Romans, and the Roman to that of their even-less-civilized successors, and knowledge of φ lay dormant until Pacioli and da Vinci revived it in the fifteenth century. It then returned to dormancy until the nineteenth century:

> The next time the golden section was consciously "rediscovered" by the artists, or more precisely the theorists of art, was around the middle of the nineteenth century, just in the time that John Taylor wrote. It cannot be said that his book was a success and it probably would have been forgotten completely if it had not been for Charles Piazzi Smyth, at that time Astronomer Royal for Scotland...
>
> At the age of forty, Smyth came across John Taylor's book and became enchanted with it. Thinking and dreaming about it, he quickly convinced himself that Taylor had barely scratched the surface. He thought and calculated and worked and, in 1864, he wrote a 600-page book called *Our Inheritance in the Great Pyramid.*
>
> It was a great success and its awed readers learned that the main item in the plan for the Great Pyramid had been nothing less than the squaring of the circle. Smyth said that the bottom square of the pyramid represented—or rather, was equal to—the circumference of a circle drawn with the height of the pyramid as its radius.
>
> Whoever read this with a critical or just an open eye should have stopped right there. Maybe the base of the pyramid measured 3055.24 feet,

> as Prof. Smyth said, but how about the height? The outer casing and with it the point of the pyramid were missing. Therefore, the height could not be measured directly. Of course if the casing had been there, one could have measured the slope angle and calculated the height from that.
>
> The slope angle must have been near 52°, so Smyth said that it originally was 51°51′14.3′ which produced a height of 486.256 feet. Therefore, the ratio of height to circumference corresponded to two *pi*. Insisting that this could not be a coincidence and apparently unaware of the fact that he had not *found* the figure but *put it in*, Smyth went on to other discoveries. [2, pp. 32, 34]

That is it, exactly: when you put something in a pyramid, it is no surprise that you can pull it out.

> At the base, one side of the pyramid measured 763.81 feet. The Egyptians naturally had not used feet as a unit of measurement; it must have been something else. Smyth divided this 763.81 feet by 365.2422 and got a unit he called the "pyramid meter." Why this figure? Obviously the builders of the pyramid had wanted to express the number of days in the year by the base line.
>
> Dividing the pyramid meter into 25 equal parts, Smyth obtained the pyramid inch, which, by a strange coincidence, differed from the English inch by just 1/1000 of an inch. Obviously the English still used the pyramid inch, but had not kept its length accurately through the millennia....
>
> Having the "pyramid meter" and "pyramid inch," Smyth really got going.

Observe that the number of days in the year was also inserted into the pyramid so that it could be taken out. There is no evidence that the Egyptians knew the length of the year to seven significant figures. Three, or at the most four, is more like it.

True pyramidologists can get around these difficulties. One method is to say that the reason there is no evidence that the ancient Egyptians knew any of the things that are so clearly in the pyramid (when you know where to look) is that they were kept secret by the possessors of the knowledge. That the possessors were good at keeping secrets is proved by how well they kept the secrets from getting out.

That is an argument to which there is no logical answer. There are only appeals to plausibility: how probable is it that some Egyptians of the third millennium B.C. calculated the length of the year to seven significant figures without telling anyone how they did it? They would also have had to keep secret the mathematics that they used, since the crude Egyptian mathematics that we know about could never have accomplished it.

The notion that the ancients knew things that we do not and kept them secret will not go away. In the Indianapolis *Star*, a newspaper of high respectability, we find printed a dispatch from the Deutsche Presse Agenthur:

> Archaeologists might be closing in on one of history's most puzzling archeological mysteries—the riddle of the Sphinx.
>
> Gypsy Graves, a leading U. S. archeologist, will return to Egypt soon with a documentary film crew to take part in the latest effort to find chambers believed hidden below the enigmatic figure in the desert near Giza....
>
> According to some accounts, a vast city exists beneath the Giza plateau, with an entrance somewhere near the Sphinx. Legends say the city contains the Halls of Record, which contain a repository of knowledge embedded in quartz crystals. Supposedly, priests sealed the chambers to preserve their secrets. [3]

Do you know of any precedents for such actions? Why would an organization shut up shop, one day, all of sudden, and disappear? Would, say, the Roman Catholic church decide to bury all its records, on quartz crystals or otherwise, and then go out of business? It is hardly likely. But of course divine intervention can get around *that*. The article goes on to give a whiff of premillennialism:

> "Explorers have been planning to do this for (more than a century), and somehow it has yet to be accomplished," Graves said. "Some people believe it will only take place when the time is ripe."

That foolishness like this can be published in the news section of a semi-major newspaper shows the appeal of the secret-doctrine idea.

But the improbable keeping of secrets is not the only way out, or around, the problem of how a people without knowledge of φ, π, or the date of the end of the world could nevertheless build them into one of their structures. It is also possible to say, and has been said, that it was not the ancient Egyptians who designed the pyramid. They may have furnished the musclepower, but the number of days in the year and so on were put there by others.

Once, in a motel, I was flipping through the many television channels provided for the entertainment of guests and came upon Charlton Heston narrating, in his most Heston-like voice, a documentary on the secrets of the pyramids. Though he avoided actually stating it flatly, he made it plain that there was good reason for believing that the pyramids had been constructed under the direction of... Atlanteans. When Atlantis sank, there were some survivors, who evidently washed up on the shores of Egypt. All of the records of the Atlanteans sank also, which is why we do not have them. Heston did not go into what the Atlanteans did after causing the pyramid to be built. To assimilate into Egyptian society seems insufficiently grand. It is more likely that they went to England to build Stonehenge. But what did they do after that? Heston did not say. Whether by accident or on purpose, he was wearing the traditional dress of the charlatan: light-colored suit, black shirt, a yellow necktie.

It is difficult to refute the argument that Atlanteans built the pyramids, though it is weakened by the lack of relics from Atlantis. But going one step further avoids even that weakness. It is to flat-out assert that supernatural beings designed the pyramid and the ancient Egyptians were unwittingly carrying out their will. There is no way to controvert this. Saying "Supernatural beings did not design the pyramid" is completely answered by "They did too." Supernatural beings can know the number of days in the year to more decimal places than we ever will, and the trillionth digit of π is at their fingertips. A proper question to ask is "*Why* did supernatural beings design the pyramid?" Pyramidologists have given answers to that one too, and we will see one of them in chapter 26.

It is much more reasonable, I think, to say that the pyramid is nothing but a huge pile of rock, designed and piled up by people. If that is so, how do we explain the close approaches to φ and π? It can be done. Let us look at the Rhind papyrus. The papyrus, written sometime around 1650 B.C., is a collection of problems and solutions, no doubt used as a text for students of mathematics, and no doubt containing much, if not all, of Egyptian mathematical knowledge of the time. It follows that it contains much, if not all or more than all, of Egyptian mathematics at the time the pyramids were being built.

The papyrus has pyramid problems in it! Here is problem number 56, taken from [5, p. 725], which was quoting from [1]:

> If a pyramid is 250 cubits high and the side of its base 360 cubits long, what is its seked?

Figure 1 is a picture of the pyramid. The "seked" of the pyramid is the lateral displacement for a rise of seven palms. (Seven palms made one cubit, and a palm was divided into four fingers.) Thus it gives a measure of the slope of the pyramid's faces. This is what Piazzi Smyth used to show that the Egyptians knew the value of π. What was being asked for in the problem was the value of d in Figure 2. That is easy to find since the two triangles are similar:

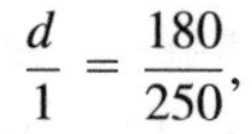

$$\frac{d}{1} = \frac{180}{250},$$

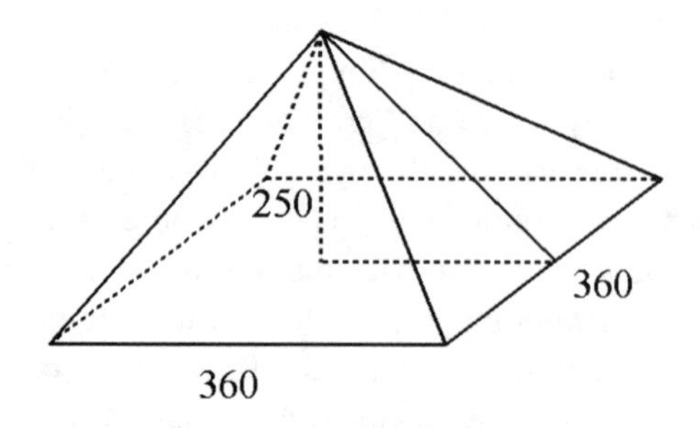

FIGURE 1 Rhind papyrus pyramid.

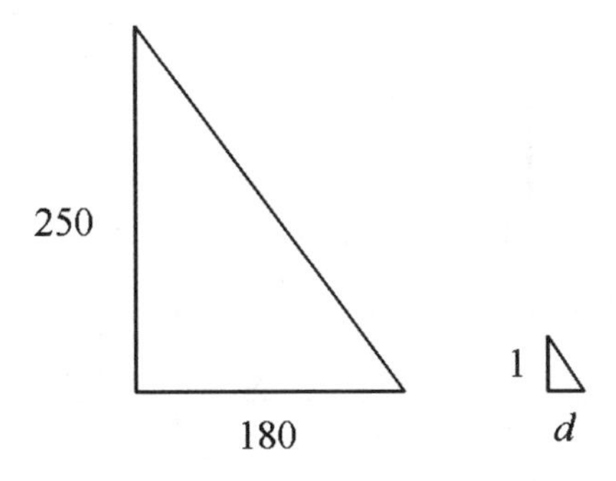

FIGURE 2 Rhind papyrus, problem 56.

so $d = 180/250 = .72$ cubits, and in palms that is $7 \times .72 = 5.04 = 5\ 1/25$. That is how we would solve the problem, but the ancient Egyptians knew nothing of decimals and could not deal with fractions like 250/180. Here is how the solution in the papyrus went:

> Take 1/2 of 360; it makes 180. Multiply 250 so as to get 180; it makes 1/2 1/5 1/50 of a cubit. A cubit is 7 palms. Multiply 7 by 1/2 1/5 1/50.
>
> | | |
> |---|---|
> | 1 | 7 |
> | 1/2 | 3 1/2 |
> | 1/5 | 1 1/3 1/15 |
> | 1/50 | 1/10 1/25 |
>
> The seked is 5 1/25 palms.

You can see that the papyrus-writer used the idea of proportion. The reason for the length of the solution is the lack of fractions other than reciprocals in Egyptian mathematics, and their clumsy way of multiplying two numbers together. Why the Egyptians never made the leap from the idea of reciprocal to the idea of fraction, which seems so natural (when you grasp the idea of one fifth, it does not seem to be a large leap to think of three of them, or three fifths) we cannot know, but they didn't. Their awkward arithmetic sufficed for their needs, and in its thousands of years of existence, no genius arose to improve it. If the Egyptians had to work this hard to solve a simple problem involving similar triangles, it does not seem likely that they could have known the value of π to seven or eight decimal places, especially since they had no notation for decimals.

In the proper pedagogical style—pose the same problem in different ways, to make sure that the student understands—the papyrus's next problem is

> If the seked of a pyramid is 5 palms 1 finger per cubit and the side of its base 140 cubits, what is its altitude?

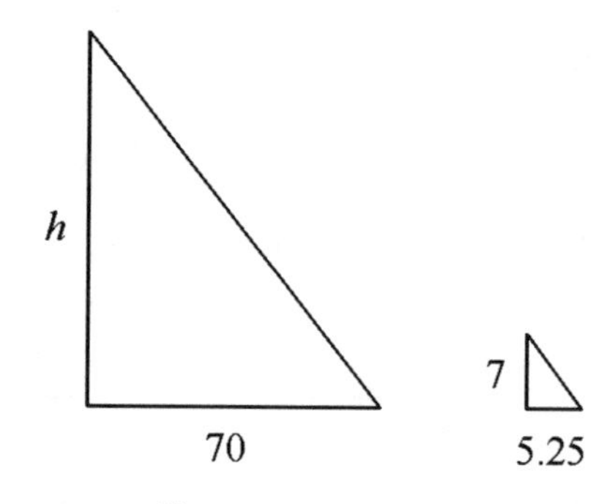

FIGURE 3 Rhind papyrus, problem 57.

As illustrated in Figure 3, this is another problem in proportion:

$$\frac{h}{70} = \frac{7}{5.25} = \frac{4}{3}; \quad h = 280/3.$$

Here is the papyrus's solution:

> Divide 1 cubit by the seked doubled, which is 10 1/2. Multiply 10 1/2 so as to get 7, for this is a cubit; 7 is 2/3 of 10 1/2. Operate on 140, which is the side of the base: 2/3 of 140 is 93 1/3. This is the altitude.

Problem 58 in the papyrus is

> If a pyramid is 93 1/3 cubits high and the side of its base 140 cubits long, what is its seked?

The alert reader of the papyrus would have seen that since this is the same pyramid as in the preceding problem, the seked is 5 1/4 palms. The papyrus was a text, and one of the purposes of a text is to allow the person teaching from it to see if students are alert.

Pedagogy apart, the Rhind papyrus has two things to tell us that are important for the pyramid. The first is that ancient Egyptian mathematics was primitive in the extreme. The second is the angle of inclination of the pyramid. The first is is not to the discredit of the Egyptians, since civilization has to start somewhere. The Egyptians were not degenerating and what they had was better than anything that anyone had before. Since no one before had much of anything beyond mere counting, it is only to be expected that Egyptian mathematics was primitive.

For one last example, here is problem 24 from the papyrus.

> A quantity and its 1/7 added together become 19. What is the quantity?

Solving that today is a problem of ninth-grade algebra. Letting, as is traditional, x stand for the unknown quantity (the ancient Egyptians never had this idea, nor did anyone else for thousands of years) the equation to be solved is

$$x + \frac{x}{7} = 19.$$

There is nothing to it:

$$\frac{8x}{7} = 19, \ 8x = 133, \ x = \frac{133}{8} = 16\frac{5}{8}.$$

Here is how the solution goes in the papyrus:

Assume 7

1	7
1/7	1
Total	8.

As many times as 8 must be multiplied to give 19, so many times 7 must be multiplied to give the required number.

1	8
*2	16
1/2	4
*1/4	2
*1/8	1
Total 2 1/4 1/8.	

The method used is that of *false position*, taught for centuries until algebra superseded it. We start by making a guess at a solution. The papyrus-writer guesses 7, not very close but the method succeeds no matter what the guess is, and finds that 7 and its seventh sum to 8. This is not large enough: we want a sum of 19, not 8. So, if we find out what times 8 gives 19, that quantity times 7 will give the answer we are looking for. Since the ancient Egyptians could multiply and divide only by 2, to find how many times 8 goes into 19 we construct the pair of columns above starting with 1 and 8. Multiplying by 2 gives 2 and 16. Multiplying by another 2 would take us past 19, so we divide to get the next line. Two more divisions by 2 give enough pieces, those marked with asterisks, to make up the total of 19. So, the first part of the solution is finished: $8(2 + 1/4 + 1/8) = 19$.

To get our answer at last, we find what 7 times 2 1/4 1/8 is by adding one times it, two times it, and four times it (any multiplication can be accomplished by doubling and addition):

1	2 1/4 1/8
2	4 1/2 1/4
4	9 1/2

Do it thus: the quantity is 16 1/2 1/8.

And there is the answer, 16 5/8, that we found before. The ancient Egyptians were clever, but very, very primitive. With no algebra, no trigonometry, hardly any geometry, and laborious arithmetic, the Egyptians were not going to make many deep mathematical discoveries and in fact they did not.

In particular, they were not going to find the value of π to very many decimal places. Problem 41 of the papyrus gives the Egyptian rule for finding the area of a circle. Since the area of a circle is π times its radius squared, this gives the Egyptian value of π. Their rule was to take the diameter of the circle, subtract one-ninth of it, and square the result. That is, if d and r stand for the circle's diameter and radius,

$$(d - d/9)^2 = \pi r^2$$

or

$$\left(\frac{8}{9}d\right)^2 = \pi\left(\frac{d}{2}\right)^2, \quad \frac{64}{81}d^2 = \pi\frac{d^2}{4}, \quad \frac{64}{81} = \frac{\pi}{4},$$

so $\pi = 256/81 = 3.1604\ldots$. ([7, p. 45] has an interesting speculation on how that value was arrived at.) The value was no doubt close enough for the uses that the Egyptians put π to—problem 41 is about the volume of a cylindrical container, the ancestor to today's silos. But it is neither advanced nor sophisticated mathematics.

An important point is that the sekeds of the pyramids in the papyrus were numbers like 5 1/25 and 5 1/4. Since writers of mathematics textbooks always want to include up-to-date applications, if only to try to keep students interested, it is likely that real pyramids had sekeds like those. It is also almost certain that the designers of the great pyramid would have fixed the angle its faces make with the ground by specifying the pyramid's seked.

As pyramid-measurers have found, the angle its faces make with the horizontal is about 52°. If we ask what seked would give an angle of inclination of about 52°, the answer leaps out. A seked of 5 1/2 palms gives an angle of inclination of 51°50.6′. That, I am convinced, is how the pyramid was built. That seked is a number simpler even than the 5 1/4 and 5 1/25 found in the Rhind papyrus—textbook writers sometimes disdain actual objects as too simple to provide proper examples—and provides a much better explanation of the pyramid angle than Piazzi Smyth's calculations, calculations that the ancient Egyptians could not have carried out.

As for φ and π, it is easy to see how those values arise, since the angle needed to get π out of the pyramid is 51°51.2′ and the angle for φ is 51°49.6′. These angles differ from 51°50.6′ hardly at all. If you go a mile horizontally, the rise up the three angles would be 6716′, 6720′, and 6723′, from smallest to largest. Four feet in a mile! Those numbers are close enough together so that someone measuring an incomplete worn-down pyramid with instruments themselves inaccurate, not to mention the human errors that observers can make, could get any one of the three angles. But the one that makes the most sense is the one that comes from a seked of 5 1/2 palms to one cubit. Of course, it does not involve mystery, secret priesthoods, lost races, or all-seeing gods. It has no romance, no wonder. It merely explains, simply and convincingly, why the rocks are piled up the way they are. How dull. But truth, and life, can be that way.

One reason why Taylor and Piazzi Smyth did not look at the Rhind papyrus and come to the same conclusion is that the papyrus was not available for them to look at. It was bought in Egypt by an English collector, Alexander Henry Rhind, and after his death in 1863 it went to the British Museum. A German translation appeared in 1877, and the official British Museum facsimile was not published until 1898.

Modern pyramidologists do not have that excuse available.

References

1. Chase, A. B. et al, editors, *The Rhind Mathematical Papyrus*, the Mathematical Association of America, vol. 1, 1927; vol. 2, 1929; reprinted, 1979.
2. Herodotus, *Account of Egypt*, translated by G. C. Macaulay.
3. Indianapolis *Star*, **92** (1994) #119 (October 2), A16.
4. Ley, Willy, *Another Look at Atlantis*, Bell, New York, 1969.
5. Midonick, Henrietta, editor, *The Treasury of Mathematics*, Philosophical Library, New York, 1965.
6. Piazzi Smyth, Charles, *Our Inheritance in the Great Pyramid*, Straham, London, 1864.
7. Robins, Gay and Charles Shute, *The Rhind Mathematical Papyrus*, British Museum Publications, London, 1987.
8. Taylor, John, *The Great Pyramid, Why Was it Built and Who Built it?* Longmans Green, London, 1859.

CHAPTER 24

Inside the Pyramid

The Great Pyramid, by Basil Stewart [1], author also of *Witness of the Great Pyramid*, *Mystery of the Great Pyramid* and other similar works, was first published in 1925, had a fourth edition in 1933 and, amazingly, is still in print. It is amazing since the book predicts the end of the world on August 20, 1953, an event that seemingly did not occur. Its main thesis is that the passageways inside the great pyramid provide a chronology of events from the beginning of the world around 4000 B.C. until the onset of the millennium.

Questions about this assertion spring to mind. One is, how did the ancient Egyptians know what was going to happen? And, if they did, why didn't they just write it down instead of going to the trouble of hauling all that rock around? The answer is that the Egyptians had no idea what they were really doing. They were being used.

> The problem of the Great Pyramid, in fact, is, like its construction, unique. Though erected *in* Egypt, it is *not of Egyptian origin*, and it is because Egyptologists, as a body, have failed to recognize this fact that they have been unable to discover its true purpose, but have supposed it to be a tomb like other pyramids, *a theory for which no evidence can be produced in support.* [1, p. 3]

The italicized phrase raises a question also. The Egyptians built many pyramids before erecting the great pyramid, and there is what seems to be a natural evolution (see Figure 1) from a funeral mound to a stepped mound, from that to a step pyramid such as the one at Sakkara (c. 2700 B.C.), and finally to the prettier and more regular shape of a pure pyramid. Since the approximations to pyramids were funeral monuments, it seems reasonable that the great pyramid was indeed a tomb, though more imposing than any that went before it. Not so:

> These earlier structures clearly indicate a period of experimental building in order to train the Egyptian workmen towards attaining the manual skill necessary in erecting the great pyramid. [1, p. 5]

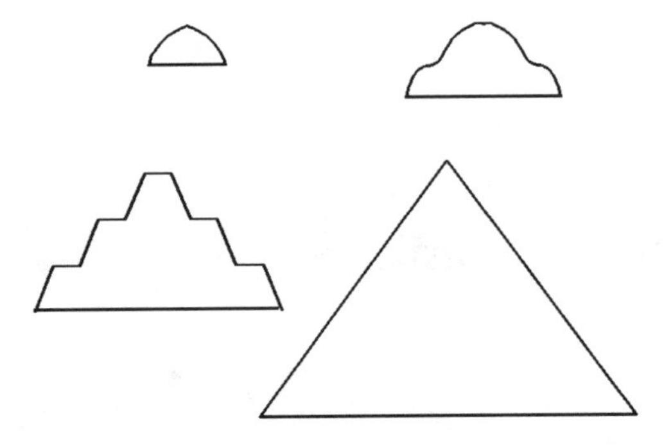

FIGURE 1 The evolution of pyramids.

The Egyptians also built pyramids after completing the great pyramid. Since the purpose of recording the future of the race had been fulfilled, these hardly seemed necessary. Mr. S. explains them by the urge to copy.

> The fact that later pyramids were [used as tombs] shows the Egyptians of that day did not themselves know the purpose the architect had in erecting the Great Pyramid, but, like opinion ever since, down to present day Egyptologists, regarded it as a mausoleum to perpetuate his name. Hence attempts made by subsequent monarchs to copy it, and their adoption of this form of burial place. [1, p. 5]

The argument that the Egyptians were unconsciously directed by higher powers to build the pyramid is unanswerable. It cannot be refuted, any more than can be the assertion that we were all created yesterday complete with all our memories. Pyramidology is thus a religion, founded squarely on faith.

The pyramid was built to send a message.

> And the message therein revealed is the same as the message of the Bible: to proclaim Christ as the Savior and Deliverer of mankind, at the same time warning us of the time and circumstances of His coming, to the end that, if we heeded it, we might be prepared for that great event.
>
> This message is given in the form of a huge graph or geometrical design, represented by the pyramid's passages and chambers, the structural changes therein defining the dates of epoch-making events in world-history. Sceptics will say, of course, that such an idea is so improbable as to be dismissed at once from further consideration. It is, however, surely just as reasonable for one person to leave a record of future history in the form of a chart, expressed geometrically in the lines of a building—a form of representation precisely analogous to the graphical methods (known as "graphic statics") employed in the solution of problems by the civil engineer—as it is for others to commit a similar record to writing, as has been done in the Bible, all acting under Divine inspiration. [1, p. 9]

It seems to me that it would have been simpler to cause the pyramid-builders to inscribe, either in a passageway deep inside the pyramid or, for that matter, on one of its faces, the message, in English, "End of the world coming September 15, 2004, 6 P.M. Greenwich mean time. Be ready!" The existence of such a message, especially if it was noted and recorded by observers before the English language had come into existence, would be truly compelling evidence. I have never seen that possibility mentioned, but no doubt Mr. S. or one of his successors would have an answer ready, or could quickly prepare one. If nothing else, there is always the reply that the ways of supernatural beings are beyond the understanding of humans.

To get the chronology, we look at the pyramid's internal passageways. Figure 2 is a rendering of Mr. S.'s more elaborate picture. There is the descending passage, intersecting the ascending passage at A. The ascending passage leads to galleries and chambers.

If we extend the lines of the face of the pyramid and the ascending passage to their intersection at O we get the origin, 4000 B.C., of the geometrical chronology. We then measure along the line OA with the scale of 1 year = 1 pyramid inch to see what has happened and what is going to happen.

The "pyramid inch" is one twenty-fifth of the pyramid cubit, and is almost identical to the English inch.

> This brings us to the first point clearly indicating the particular significance of the Great Pyramid for the Anglo-Saxon race, for our inch varies from the Pyramid inch by only a thousandth part (1,000 Pyramid inches equals 1,001 British inches), while originally they were identical, this minute difference

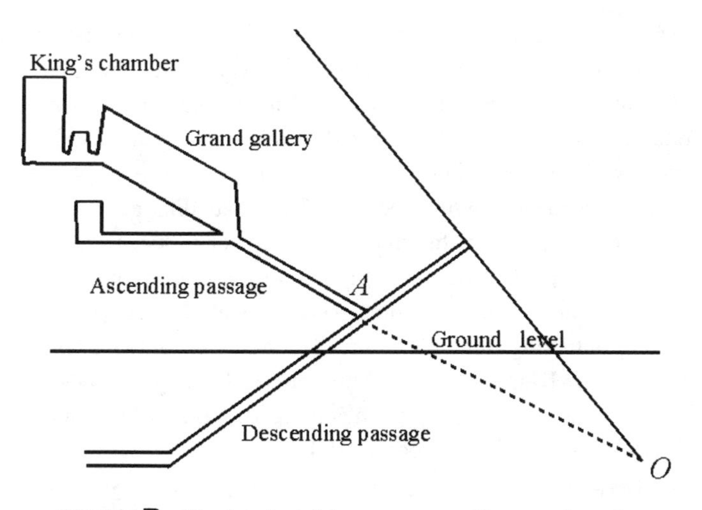

FIGURE 2 The interior of the great pyramid, approximately.

being due to lapse of time, and to the fact that for long periods we had no proper standards of length for comparison.

Had it not been, therefore, for the preservation by the English-speaking peoples of their inherited measure—the inch—which has not only survived from remotest antiquity, but has defeated the various attempts that have been made from time to time to displace it by the unscientific metre, the meaning of the Great Pyramid would never have been revealed, for otherwise the key to it would have been lost, and this witness in the land of Egypt would never have been unlocked. It is not surprising, therefore, that all who have been instrumental in deciphering it have belonged to the Anglo-Saxon race.

[1, pp. 13–14]

To the question of how the inch got from Egypt to England, the answer is that the English are descendants of the ten lost tribes of Israel, who migrated from the eastern Mediterranean to the chill and foggy isles of Albion long, long ago, leaving no record of their travels but taking their inches with them. This explains the connection between the pyramid and another collection of stones, Stonehenge.

Stonehenge, the Eastern origin of which is established by archaeology, folklore and tradition, is set out on exactly the same year-circle, proving that its builders were descended from the same race, and used the same system of metrology as the builders of the Great Pyramid. A circle of 3,652.42 pyramid inches circumference, equivalent to the number of days in a solar year on a scale of ten inches to a day, falls precisely internal to the outer ring of stones forming the circle of Stonehenge. This fact proves that the astronomical conceptions of the ancient British megalithic builders originated in Egypt, where they reached their highest constructive expression in the erection of the Great Pyramid. [1, p. 15]

Leaving Stonehenge, and the question of whether its ancient builders knew the length of the year to six significant figures or were merely inspired to incorporate that number into their structures without understanding its significance, let us return, with a ruler, to the pyramid. When we measure along *OA* to its intersection with the descending passage, we have arrived at 1486 B.C., the date of the exodus from Egypt. The end of the ascending passage, or almost the end—where it intersects the floor (extended) of the "Queen's (or Jew's) Chamber"—dates, by the pyramid-inch ruler, to 4 B.C., the year of the Nativity. The end of the roof of the ascending passage is at 30 A.D., the date of the Crucifixion. Continuing along the Grand Gallery, we come to the Great Step at the entrance to the King's Chamber. This dates to January 25, 1844. You may not recall anything important happening on that date or, for that matter, on any date in 1844—but something important did:

The next date defined is that marked by the Great Step—January 25th, 1844—one of immense importance astronomically in the Pyramid's representation

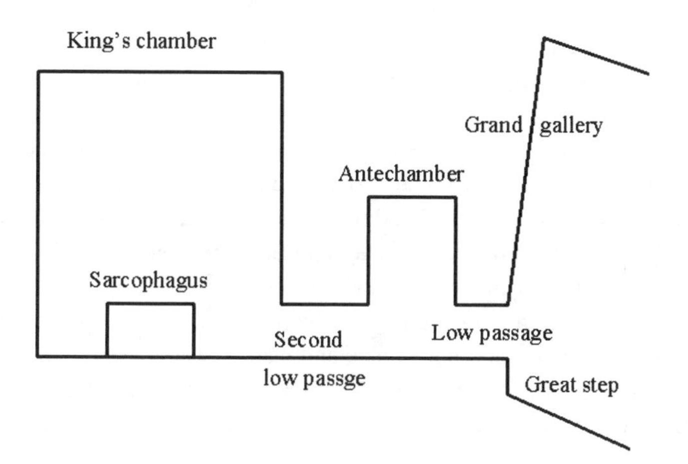

FIGURE 3 The ante-chamber and the king's chamber.

> of chronology, the explanation of which is given in our larger work, but is somewhat too technical and involved to explain in this introductory essay. [1, p. 32]

Now let us step up the Great Step and onto the level. See Figure 3. The time scale now changes to one inch per month instead of one inch per year.

> The breaking of continuity of the Gallery floor at this point, and the change of direction to the horizontal, implies a change of chronologic scale... the rate for this level portion being an inch to a *month*; not our varying month of thirty or thirty-one days (since a varying unit cannot be geometrically represented by a fixed one—an inch), but the ancient Egyptian and Biblical month of thirty days. [1, p. 32]

There is an element of arbitrariness in this. Mr. S. tries to allay any such feeling.

> The Pyramid's passage system, in fact, is analogous to the working drawings of some engineering undertaking. There is the general drawing showing the whole scheme on a small scale, while separate portions are given on an enlarged scale to show the details of construction for the contractor to carry them out.
>
> Not only is the scale thus enlarged to show the greater detail of modern history, but there is also a clear distinction involved between these two representations. As already pointed out, the message of the Great Pyramid is specifically addressed to modern times. That is to say, its general, small-scale system was intended to be studied, and its significance understood with respect to past history, during the years defined by its special, larger-scale system—the opening years of the twentieth century. And such precisely has been the case. [1, p. 33]

Yes, precisely, because the beginning and end of the Low Passage have dates, on the new enlarged scale, of August 5, 1914 and November 11, 1918, the dates, to the day, of England's entrance into and exit from World War I. There is some room for doubt about such accuracy, since even in the enlarged scale one day would take up a span of only one thirtieth of an inch and I cannot imagine that the beginning and ending of the Low Passage were so sharply marked that they could be located that precisely. The conditions of taking the measurements make it unlikely. Crawling around inside a pyramid, through passages four feet high or less, dimly lit, dusty, suffocating: imagine having to carry along steel tapes and notebooks, and take measurements. (I have been inside a pyramid—not Cheops', but Khefren's—and I did not like it one bit. Any measurements that I took would not have been accurate to one-thirtieth of an inch.)

But some pyramid-measurers can do even better than one thirtieth of an inch.

> This enlarged scale is defined in the Pyramid itself. From the center of the Ante-Chamber to the south wall of the King's Chamber is a distance of 365.242 inches, equivalent to the number of days in a solar year. This distance is the circumferential measure of a circle whose diameter equals the length of the Ante-Chamber (116.26 inches). [1, pp. 32–33]

A measurement of 116.26 inches: accuracy to within one one-hundredth of an inch; 365.242 inches: accuracy to one *one-thousandth* of an inch! Is it possible that Mr. S. was serious? Surely he must have known that such accuracy was absurd. But if he did, that would imply that he was using his measurements cynically, to mislead those to whom numbers meant little or nothing. I doubt that. The power of delusion over the human mind is so great that I think it was possible—likely, even—that he believed in his numbers, every single significant digit of them.

Although the pyramid could give the date of Armistice Day with 100% accuracy, it slipped a bit later on. Clearly, the start of the second Low Passage (see Figure 3) must have some significance.

> The open space of the Ante-Chamber represents a period of truce in tribulation.... This truce ended on May 29th–30th, 1928, when the Second Low Passage was entered, representing another period of trial for the "builders." [1, p. 34]

The "builders" are the "Anglo-Saxon race" whose forebears were responsible for the construction (though *not* the design) of the pyramid. Mr. S. was not able to put his finger on anything as specific as the Armistice for May 29–30, 1928. The best he could do:

> Few today will dispute the fact that, since the summer of 1928, we have been passing through an extremely critical period, in some respects even more critical than at the crisis of the late war, and it was on May 29th, 1928, that the various factors, economic and industrial, which have developed into this crisis, now involving the whole world, first began to reveal themselves. This crisis is indicated as continuing till September 15th, 1936. [1, p. 35]

Weak, quite weak. The world is *always* in a crisis, it is *always* in a critical state, and events are *always* unprecedented. A glance at any newspaper will verify this. Every generation sees itself as in the midst of change at a rate unparalleled in all of history, facing dangers larger and more menacing than have ever been seen before, and being closer to the brink than at any time in the past. Previous generations may have thought that *they* experienced crisis, but they were wrong. *They* lived in an age of innocence (the age of innocence is always about seventy-five years in the past) and they innocently thought that *they* were experiencing rapid change and danger, but they were wrong. *We* are right, we assert. We are not right, we are merely self-centered.

Mr. S. presents more evidence that he was living in the last days:

> There is the following remarkable fact respecting the two low passages symbolizing periods of stress. Their combined length is exactly 153 inches—the number of fishes Peter drew to land in the miraculous haul in the sea of Tiberias (John xxi, 11). In Matt. xiii, 47–9, is a parable giving this incident in parable form, wherein Christ likens the good fish, gathered safely to land, to the just, whom the angels separate from the wicked, who are the bad fish cast back into the sea—ending with the words: "So shall it be at the end of the age." The number 153 is thus here connected with the "elect," for whose sake these days are shortened (insertion of Ante-Chamber), and identifies the Great Pyramid's symbolic prophesy with our Lord's prophesy concerning the final tribulation which was to precede His Second Coming.
> [1, pp. 35–36]

He did not note that 153 is a triangular number, and hence just right to appear in a pyramid with triangular faces.

Mr. S., writing in 1925, knew that many readers of his book would experience September 15, 1936 and he prepared them for the possibility that it would not be marked by earthquakes, tornados, and the parting of the heavens to the blare of trumpets.

> Our Lord also has told us that, at the time of the end, mankind would be warned by previous events as to the nature of future events. This is precisely the sequence of warning and future effect given by the Pyramid's Ante-Chamber. It likewise uttered the warning "Watch!"
>
> Had this warning been heeded, we should not have had the exaggerated, and often wild, prognostications made in the popular press by irresponsible journalists and others about the Great Pyramid and May 28th, 1928, which

> were naturally seized upon to discredit the whole subject. Time, however, quickly proved the folly of jumping to conclusions, a step the writer, for one, always warned against. It is to be hoped that similar foolishness will not be indulged in as we approach the Pyramid's next outstanding date—Sept. 15th, 1936—but wait till it arrives. Also, do not necessarily expect something spectacular, and remember its significance may not appear till it has passed. [1, pp. 41–42]

In fact, nothing spectacular occurred. Similarly, we know that nothing spectacular happened on August 20, 1953 either, though that marks the end of the king's chamber. Mr. S. said that we may have to wait a little longer:

> As already pointed out, this enlarged chronologic system represented by the two low passages, ante-chamber and King's Chamber, constitutes the "working drawings" of the British Race as the Great Architect's builders on earth. This is shown by the fact that they are set out in our unit of measure—the inch, and August 20th, 1953, is the limiting date for their application, though the contract itself must be completed by September 15th, 1936, the final date of the great "Seven Times"—but *may* be completed sooner. That is to say, responsibility is taken out of our hands after August, 1953, and vested solely in the Architect.... That is to say, actual work of world-reconstruction does not begin till after August 20th, 1953, and is completed by A.D. 2001, when the Millennium sets in.... The symbolism indicates that the dominance of the present evil forces in the world—the "unclean spirits" of Rev. xvi, 18—is ended by September 16th, 1936, and mankind is completely purged of their power by August 20th, 1953. The period of forty-eight years thence to A.D. 2001 denotes a second and more advanced stage of the cleansing, restoration, and reorganisation—spiritually, morally and physically—which the whole world will have to undergo in order to fit it for the ideal conditions of Christ's Millennial reign. [1, pp. 36–37]

Such are the dangers of being specific when prophesying. I do not know how, if at all, the pyramid-measurers got around the failure of the 1953 date. Perhaps the cleansing, and so on, is actually taking place at this very moment, but we are too close to it to see it. In 2001, now almost here, all will be made plain.

What I think will be made plain as 2001 slides into 2002 with no signs and portents in the heavens is that many millennarists were misinformed about the date of the end of the world. What will also be made plain is that the millennarist movement will not therefore be killed. Revisions will be made. The idea that our generation is special and it will see the end is, though wrong, so appealing that it will no doubt never be stamped out until the end, in one form or another, actually arrives.

What all this pyramid-measuring illustrates is a sort of numerology-in-reverse. Numerology is based on the Pythagorean assumption that all is number: that numbers rule and reality behaves in accordance with them. The pyramid-measurers would like us to believe that the numbers they have found in the pyramid are telling us what will happen. But what Mr. S. and his successors are actually doing is taking an already-formed conclusion and searching for numbers that will support it. The numbers are not telling them what to do, they are telling the numbers what to do. But numbers, being powerful, have the last laugh. No one can tell numbers what to do.

Reference

1. Stewart, Basil, *The Great Pyramid*, 1925, fourth edition 1933, reprinted by Sun Books, Santa Fe, New Mexico, 1992.

CHAPTER 25

The Pyramid, Stonehenge, the Malaysian Lottery, and the Washington Monument

This chapter contains no new ideas. However, it does have new examples that illustrate old ones.

A sign of intelligence is the ability to see connections between things that the less acute miss. What then shall we call the ability to see connections between things that have no connection? Bonnie Gaunt, the author of *The Magnificent Numbers of the Great Pyramid and Stonehenge* [2], could see connections between the great pyramid of Egypt, Stonehenge, and the Bible, as well as one between π and the length of the year. We will go into the first set of connections in a moment. For the second [2, p. 76], note first that the angle whose sine is $\sqrt{\pi}/4$ is $26^\circ 18.9' 7''$, and this is precisely the angle, Ms. G. says, of the passageways inside the great pyramid. A right triangle with a hypotenuse twice the length of the king's chamber in the pyramid (2×412.13168792 pyramid inches, she says, all digits presumably significant) will have, opposite that angle, an altitude with length 365.242. I have no idea where she got that length to eleven significant figures, but the chamber's width, 206.0658439604 pyramid inches, has thirteen.

The idea of her book is that numbers have power—in this case, the power to convince. Her foreword begins:

> Numbers have long been a source of fascination and wonder to man. The ancient Pythagoras was so awed by the relationship of numbers to time and space that he theorized: "Numbers are the language of the universe." And indeed Pythagoras was right....
>
> The discovery of the inter-relationships of the numbers involved in the geometry of the Great Pyramid and Stonehenge led to a comprehensive investigation into the "language of the universe," and a glimpse into the very

> foundation of creation. The numbers reveal the evidence of the existence of an intelligent Creator. [2, p. vii]

Her goal is to show that the pyramid, Stonehenge, and gematria provide Christian evidence so strong as to be irresistible.

More than most pyramidologists, Ms. G. strives for eloquence:

> On the rocky plateau of Gizeh, fifteen miles from Cairo, stands the worlds most amazing wonder, the Great Pyramid. It stands silent and serene against the Egyptian sky, yet it stands in bold defiance against time: it stands as a sacred memorial to the intelligence of its builders, and to their knowledge of time, space, and the universe. [2, p. 3]

After an account of the penetration of the pyramid by the Arabs in 820 A.D., she comes to the discovery of the casing stones that allow its slope to be determined:

> In 1836 Colonel Howard-Vyse undertook the tumultuous task of clearing away the rubble of limestone, sand and debris that had accumulated through the centuries. It was a foreboding task, as some of the piles were as much as fifty feet high. [2, p. 12]

I think that Ms. G. was searching for "forbidding" when she hit on "foreboding" and that "tumultuous" should probably be "tremendous" or something similar. In any event, he measured the angle and

> He found the angle to be $51° 51' 14.3''$. [2, p. 1]

That is, he found the precise angle that Piazzi Smyth needed to get both π and the number of days in the year out of the pyramid. As I tried to convince you in chapter 24, the pyramid was built so that its faces had a slope of $5\frac{1}{2}$ palms to the cubit, that is, $5\frac{1}{2}$ palms to 7 palms, or 11 to 14 (see Figure 1). The angle corresponding to that slope is quite close to Piazzi Smith's $51° 51' 14.3''$—it is $51° 50' 34''$. It would not be hard to mistake one angle for the other.

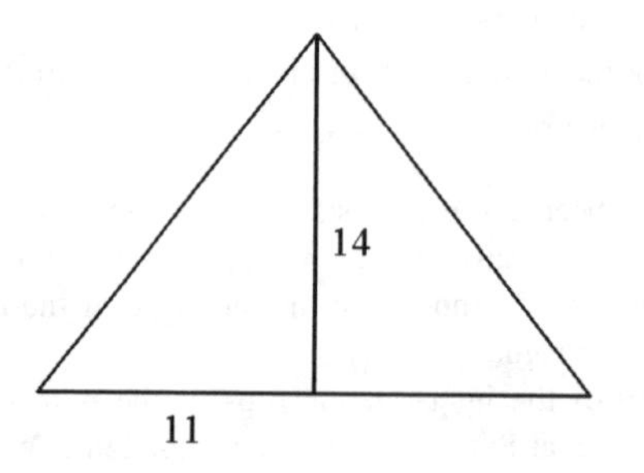

FIGURE 1 Slope of the great pyramid.

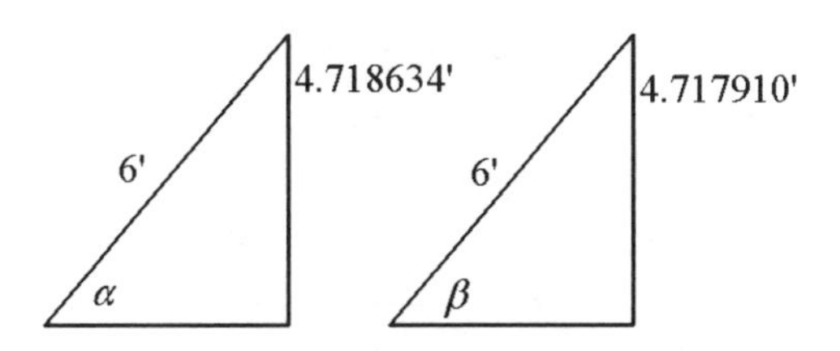

FIGURE 2 Which looks like the pyramid?

In her book, Ms. G. has a picture of a casing stone with a berobed Egyptian standing stolidly beside it [2, p. 14]. It looks as if its hypotenuse is about six feet long. It looks a little worn, too, as would you or I after four thousand years. Disregarding wear, if its hypotenuse was six feet long, then the difference in heights for the two angles, $\alpha = 51^\circ\ 51'\ 14.3''$ and $\beta = 51^\circ\ 50'\ 34.0''$ is .000724 feet, or .00868 inches (see Figure 2).

It is quite impossible to measure distances from one uneven piece of rock to another with that much precision. To report the angle as being anything more precise than "about 51° 50′" would be saying too much and "near 52°" would be more honest. But $51^\circ\ 51'\ 14.3''$ precisely is what it has to be, so that it what it is. Ms. G. says,

> The magnitude of this discovery is tremendous. It has opened for us a world of knowledge and insight into the origin and purpose of the Great Pyramid and into the astounding knowledge and understanding of its builders; and we are faced with the realization that ancient man, our forefathers 4,000 years removed, were not the ignorant barbarians that we have been taught to believe. They were intelligent men, with a knowledge of the universe in which they lived.
>
> Science and astronomy today can give us the size relationship of the earth and its moon. Did man, 4,000 years ago, know this relationship? If we were to place the circle of the moon tangent to the circle of the earth (as in Figure 6), the exact proportions of the Great Pyramid can be constructed on their combined radii, showing the casing stone angle of 51° 51′. Did the builders of the Pyramid know the size of the earth and its moon? How could they? [2, p. 13]

Ms. G.'s Figure 6 is equivalent to Figure 3. Note that she does not cite the angle θ as being the pyramidologist's $51^\circ\ 51'\ 14.3''$. That is because it is not. Since

$$(3960 + 1080)/3960 = 5040/3960 = 14/11,$$

what Ms. G. has, without knowing it, is the *real* angle of the pyramid, slope $5\frac{1}{2}$ palms to the cubit.

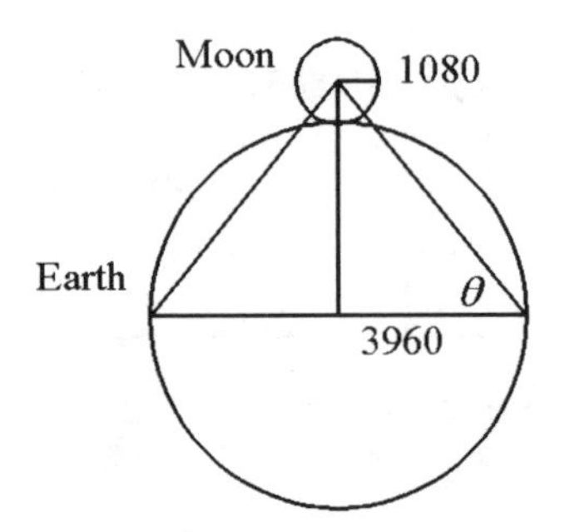

FIGURE 3 Earth and moon touching.

However, this does not show that the pyramid architects knew about the diameters of the sun and moon. It is yet another example of the wide applicability of the Law of Small Numbers. The author chose rounded-off values for her radii. 504 and 396 are small numbers and, what is more, small *round* numbers with many divisors ($504 = 2^3 \cdot 3^2 \cdot 7, 396 = 2^2 \cdot 3^2 \cdot 11$). As we know, there are not enough small numbers for the many demands made on them, nor enough round numbers.

You may nevertheless have some residual astonishment: isn't it amazing all the same that there is that 14-to-11 ratio? No, it is not. It may have been that putting the moon on top of the earth was the very first thing that Ms. G. tried, but it also may not have been. In fact, it is not likely that it was. The sun and the earth are more fundamental than the moon and the earth and it would be natural to begin by looking at various distances and diameters involving them, trying to find something close to $52°$. There may have been many previous diagrams, many tries before the inexorable operation of the Law of Small Numbers produced a success. And, if Ms. G. had not hit on her $14/11$ with the earth and the moon, there could have been other diagrams, perhaps with diameters of the orbits of the heavenly bodies instead of their radii. There are many, many numbers to choose from and even many more ways to arrange them. When the numbers are small, the Law of Small Numbers will produce results such as Ms. G. found. It is the Law and it cannot be broken.

Ms. G. now goes north.

> On the lonely wind-swept Salisbury Plain in the southern part of England stands another ancient stone monument, Stonehenge. A 4,000 year old mystery still shrouds those cold and lonely stones like the mists that hang heavily over the plain. [2, p. 17]

This is not the place for a description of Stonehenge. It was built, in stages, in the third millennium B.C. by people who left no record of how or why they did it. Suffice it to say that it has structure, a *lot* of structure. There are many rocks: there is the Bluestone Horseshoe, the Trilithon Horseshoe, the Sarsen Circle, the Bluestone Circle; there is the Altar Stone, the Slaughter Stone, Station Stones, and the Heel Stone. (I am taking Ms. G.'s word for all of this. Even if a detail or two is controversial or even wrong, there remain plenty of rocks.) In addition there are holes: Stone Holes, Post Holes, "Z" holes, "Y" holes, and fifty-six Aubrey Holes. If you are looking for numbers, you have an abundance of places to find them.

Ms. G.'s reason for bringing up Stonehenge is that, she says, it and the pyramid were closely related.

> The next obvious question is, of course, what has Stonehenge to do with the Great Pyramid? [2, p. 23]

The obvious answer is, "Nothing." That two groups of humans, far separated in space, should both have the idea of building stone monuments so that later ages should marvel is not at all surprising. (Just as Egypt contains many pyramids, the British Isles and Brittany contain a variety of henges, some with stones in circles and some with them in ellipses, and there even was a wooden henge near Stonehenge, discovered in an aerial photograph.) The Mayans built pyramids. The impulse toward monuments seems to be part of our racial heritage, as is the urge to incite wonder and achieve immortality, even if only through stone, and even if the effort is vain. "My name is Ozymandias, king of kings,/ Look on my works, ye mighty, and despair!" (P. B. Shelley, 1817) springs to mind when stone monuments are mentioned.

But to certain minds, events do not happen independently of each other. Nothing occurs by chance. This attitude is held both by occultists and children. Children, having volition themselves, see it everywhere. It is a natural and charming projection. The child has wants, sometimes very large and urgent wants, so what more natural than to suppose that animals have similar wants, as do trees, and rocks? Egocentricity and lack of experience lead children to assume that all the world is like themselves. If all things have will, then everything occurs for a reason.

Children grow up and discard the idea that the chair tripped them because it was malicious, but the idea that events can happen at random, for no reason at all, is a hard one to grasp completely. Further, it is an idea that many people do not want to grasp. Randomness is very close to chaos, and life is, among other things, a struggle for order and against chaos. Some people flatly refuse to admit that chance exists.

Evidence for this comes from [3], an article in an English-language newspaper published in Petaling Jaya, Malaysia. The article is a regular feature, one entire page, devoted to giving lottery advice. The lotteries in question—there are several—reward those who foresee what four-digit integer will be drawn, presumably at random. Here is an excerpt from the article:

> The five leaders here are series "8000", "9000", "5000", "4000", and "3000". They are very close.
>
> My choice here is number **8512**. It won a special prize on July 4.
>
> I have a strong feeling that it will be like number 5802, which has already won twice. Of course, number 5802 can score a third win. Nothing is impossible in this four-digit game.
>
> In this game, for your guidance, numbers starting with "87..." are doing well—30 prizes so far. More looks likely on the way.
>
> My second fancy is number **9218**. It has not won yet. But numbers 9213/9215/9216 have already done so.
>
> Series "9000" has taken a big jump to second place and it is still going great guns.
>
> No doubt it will give the other series a great fight in the last five months of the year.
>
> Please see statistics below.

1 + 3D
(After draw on July 25)

Position	Series	Prizes Won
1	8000	221
2	9000	218
3	4000	217
4	5000	216
5	3000	210
6	1000	206
7	7000	206
8	0000	198
9	6000	195
10	2000	180

It may be that the writer knows that lottery numbers come up at random and that all the talk of the 9000s going great guns is complete foolishness. Or—and I think this more likely—he believes every word he writes. In either case, what he writes is read and no doubt taken seriously by many readers of the paper. Otherwise, it wouldn't be published.

Actually, the distribution of winners is about what would be expected. The value of χ^2 is 7.073, and the fiftieth percentile of the χ^2 density with 9 degrees of freedom is 8.343, so the numbers are coming up with a little more evenness than average, but not so much as to make any statistician suspicious.

The reason for this dead-serious page of nonsense is our reluctance to admit chance into our universe, and our sense of the power of numbers: they have such power that they are *alive*. They have existence and will; they *want* to come up, they are in a race, they struggle to appear, and some succeed better than others. Also, they have memories, good ones, and they know what their neighbors are doing. Writing about a different lottery, the writer said

> My second fancy (not fence) is number **8958**. It has still to win. There are two reasons to support my choice.
>
> One, numbers starting with "89..." are doing well—they have won 25 prizes up to July 23.
>
> Numbers 8988/8953/8950/8978/8938/8968/8948 and 8956 have won. Therefore 8958 has bright winning chances.

The writer may think that the numbers actually *are* alive. Some newspapers in the United States print lottery statistics, but I think that they stop short of regarding the numbers as horses in a race. They leave that inference to those of their readers who want to make it.

The lottery writer's view of the world is close to the occult and mystical idea that All is One. If All is One, then everything is connected to everything else and there are no independent events. Each person, each object, is embedded in the all-encompassing... ether? jelly?... in the substance of the All-One and anything that happens to anything makes... the ether vibrate? the jelly quiver?... makes the substance react. As is usual in mysticism, words fail. But lottery numbers mystically affect each other and, for mystical reasons, Stonehenge and the pyramid were not, and could not have been, independent.

Thus, we should not be surprised when Ms. G. writes,

> If we were to start walking from Station Stone 93 to Station Stone 91, and keep walking in that same direction, we would eventually bump into the Great Pyramid. In other words, the exact azimuth from Station Stone 93 to Station Stone 91 is, in fact, the arc of a great circle that passes through both Stonehenge and the Great Pyramid. The architect of Stonehenge has left for us a road map, if you please, pointing directly toward the Great Pyramid.
>
> [2, pp. 24–25]

We can take that exactness with several grains of salt. The stones are not points, so there is quite a bit of play in finding the direction determined by them. (Ms. G., who sounds as if she has seen them, says [2, p. 26] that they are "rough, unhewn stones.") I have no idea how large the stones are, or how far apart they are, but let us suppose that they have radii of 6 inches and are 200 feet away from each other. Those are fairly needle-like stones, and I think that the diameter of Stonehenge is less than 200 feet. Whatever the numbers are, the conclusion is the same.

As Figure 4 shows, lining up two stones could produce an error as large as $\theta/2$, and $\theta = 17'11''$. Of course the proper direction would be the one determined by the line going through the center of both stones, but measurements

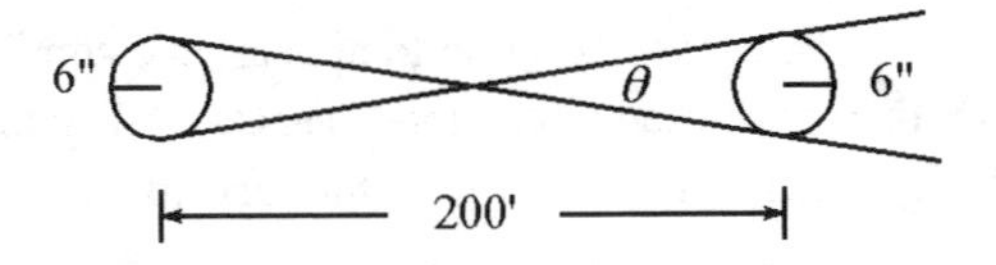

FIGURE 4 Stonehenge stone sighting.

are hard to make precisely and, if you were looking for a pair of stones to line up in a direction that you knew already, the temptation to err a little in the right direction, perhaps unconsciously, might be given into. There exist many examples of measurers who, wanting a measurement to come out a certain way, found that it did, even when it didn't. There is even the story of the pyramid-measurer who was caught filing down a stone, so his measuring would give the proper result. Though Ms. G. probably did no filing and the two stones *might* have been lined up on the pyramid, they might have missed by a bit. Even if the error was only $4'$, over the distance from Stonehenge to the pyramid it would amount to quite a few miles.

You may say, "So what's a mile or two? They line up *almost* exactly. That can't be a coincidence!" Yes, it can. Think of the number of stones at Stonehenge. Think of all the directions determined by pairs of them. If there are n stones, there are $\binom{n}{2} = n(n-1)/2$ different directions. If $n = 60$, and there are many more than 60 stones available at Stonehenge to be sighted along, that is 1,770 directions. That is, on the average, almost 5 directions for each angle from $0°$ to $360°$. It would not be astonishing that one of those 1,770 directions would point straight at the pyramid, given that the stones are not points and determine not directions but, as in the figure, ranges of directions.

Ms. G. is just warming up. She now proceeds to introduce quite a few units. We of course have the **pyramid inch**, 1.00106 ordinary inches. Then there is the **sacred cubit**, 25 pyramid inches. The **reed** is such that the base of the pyramid has perimeter 288 reeds, so one reed is 10.65 feet. The **royal cubit** is 1.7203907 feet. The **remen** is the side of a square whose diagonal is one royal cubit, so a remen is $1/\sqrt{2}$ of a royal cubit. The **Sumerian cubit** measures 1.65 feet. The **great cubit** is one-sixth of a reed. The **megalithic yard** (we have left Egypt and gone to Stonehenge) is 2.72 feet and the **megalithic mile** is 14,400 feet.

We can now start to make discoveries [2, pp. 41, 64]:

Diameter of a square containing the circle of the earth	31,680 miles
Perimeter of a square containing the circle of the moon	3,168 megalithic miles

Diameter of the sun	316,800 megalithic miles
Mean circumference of the Sarsen Circle	316.8 feet
Radius of earth	31,680 furlongs
Long side of Station Stone rectangle	3168 inches

The amazingness of all this should be tempered by noticing the variety of units used. Also, many of the numbers manipulated to give 3168 or a multiple thereof are not definite. For example, the diameter of the sun could be assigned many different values, depending not only on how accurate your measuring device was, but by what you meant by *diameter*. The sun is not a solid sphere and, were you in its near vicinity and able to keep sufficiently cool, you would have a difficult time in deciding where to draw the line between sun and not-sun. Similarly, the "circle of the earth" is not a circle since the earth is not a sphere, so the earth can have many different squares, all with different perimeters. When you combine this with the multiplicity of units to measure with—the six in bold above, along with the inch, foot, yard, furlong, and mile, all of which the author uses in one place or another—it is less amazing that some numbers, or their multiples by ten, can be made to appear.

Ms. G. combines measurements of length and area with gematria, which is why 3168 is highly significant. The numerical value of "Lord Jesus Christ" is 3168:

Κυριος Ιησους Χριστος
20 + 400 + 100 + 10 + 70 + 200 = 800
10 + 8 + 200 + 70 + 400 + 200 = 888
600 + 100 + 10 + 200 + 300 + 70 + 200 = 1480
800 + 888 + 1480 = 3168

"Well," you may say, "maybe some of those 3168s were a little phony, but how can you explain *that*? That can't be coincidence!" Yes, it can. At the risk of repeating myself and of pointing out the obvious, Ms. G. could have done one of two things. She could have first compiled all those 3168s and then looked for some way of referring to God, Christ, or the Holy Ghost—and heaven knows there is no shortage of titles and appellations from which to choose—in Greek, Latin, or English that would, using some alphabet, gematrize to 3168. It would be work, but I have no doubt that it could be done. Or, she could have started with the title and looked for physical things, astronomical, pyrimidical, or Stonehengical, that could have 3168, or a multiple thereof, wrung out of them. If you start with a large catalog of numbers, each translated into a dozen or so units, that would be easy. Compiling the catalog would be work, but I have no doubt that it also could be done.

It was Ms. G.'s idea that God caused the pyramid and Stonehenge to be built as they are, with the purpose of convincing us that he or she exists. All those coincidences cannot be coincidence, she says, there must be something behind them, and who else could it be?

Here are a few more examples of what Ms. G. thinks is proof of the divine origin of the pyramid and Stonehenge but which are actually empirical verifications of the Law of Small Numbers:

256—Truth (αληθης)
.0256 of a square furlong—area of a square drawn on outer face of Sarsen Circle
.256 of an acre—area of a square drawn on outer face of Sarsen Circle
256 feet—distance from Sarsen Center to base of Heel Stone
256 Sumerian cubits—perimeter of a square drawn on outer face of Sarsen Circle

281—Lamb (αρνιον)
28.1 megalithic yards—one side of a square of same perimeter as inner face of Sarsen Circle
.0000281 of a square mile are of a square with the same perimeter as inner face of Sarsen Circle

87—Everlasting Father (אביצד)
87 remens—diameter of outer face of Sarsen Circle
8.7 square reeds—area of a square of same perimeter as Bluestone Horseshoe
87 Sumerian cubits—radius of Aubrey Circle [2, pp. 112–115]

And so on. An enormous amount of work has been done here, taking what must have been hundreds of lengths and areas, expressing them all in a dozen or so different units, and finding matches. But the work was done. For example [2, p. 133], the long side of the rectangle formed by Station Stones 91, 92, 93, and 94 measures

3,168 inches
264 feet
88 yards
.4 of a furlong
.5 of a mile
217 remens
160 Sumerian cubits
50 great cubits
25 reeds

Meters are evidently taboo, probably because they were so obviously constructed by people, but why Ms. G. does not go on to light-years, parsecs, and other measures I do not know. ".5 of a mile" is a misprint for ".05 of a mile."

Yet more work was necessary to do the further matching with the gematrized words or phrases from the Bible:

1408, Savior (Greek)
352, The Way (Greek)
2288, Christ the Lord (Hebrew)
176, Counsellor (Hebrew)
1056, "the joy of thy salvation" (Hebrew)
555, the face of the sanctuary (Hebrew)
3330, God the Saviour (Greek)
291, Earth (Hebrew)

and so on. It was work that was bound to pay off, as guaranteed by the Law of Small Numbers.

Martin Gardner showed how easy it is to apply the Law:

> Both Taylor and Smyth made a great deal of the fact that the number five is a key number in Pyramid construction. It has five corners and five sides. The Pyramid inch is one-fifth of one-fifth of a cubit. And so on. Joseph Seiss, one of Smyth's disciples, puts it as follows: "This intense *fiveness* could not have been accidental, and likewise corresponds with the arrangements of God, both in nature and revelation. Note the fiveness of termination to each limb of the human body, the five senses, the five books of Moses, the twice five precepts of the Decalogue."
>
> Just for fun, if one looks up the facts about the Washington Monument in the World Almanac, he will find considerable fiveness. Its height is 555 feet and 5 inches. The base is 55 feet square, and the windows are set at 500 feet from the base. If the base is multiplied by 60 (or five times the number of months in a year) it gives 3,300, which is the exact weight of the capstone in pounds. Also, the word "Washington" has exactly ten letters (two times five). And if the weight of the capstone is multiplied by the base, the result is 181,500—a fairly close approximation of the speed of light in miles per second. If the base is measured with a "Monument foot," which is slightly shorter than the standard foot, its side comes to $56\frac{1}{2}$ feet. This times 33,000 yields a figure even closer to the speed of light.
>
> And is it not significant that the Monument is in the form of an obelisk—an ancient Egyptian structure? Or that a picture of the Great Pyramid appears on a dollar bill, on the side opposite *Washington's* portrait? Moreover, the decision to print the Pyramid (i. e., the reverse side of the United States seal) on dollar bills was announced by the Secretary of the Treasury on June 15, 1935—both date and year being multiples of five. And are there not exactly twenty-five letters (five times five) in the title, "The Secretary of the Treasury"?
>
> It should take the average mathematician about fifty-five minutes to discover the above "truths," working only with the meager figures provided by the *Almanac*. Considering that Smyth made his own measurements, obtaining hundreds of lengths with which to work, and that he spent twenty years mulling over these figures, it is not hard to see how he achieved such remarkable results. [1, pp. 178–179]

Most of Ms. G.'s 200 pages are devoted to the piling up of more and more numbers, designed to convince the reader that they could not have been accidental and must have been planned by a divine intelligence. She does not at all go into the question of why God should create puzzles as intricate as this one (of many that could be quoted), about the empty coffin in the king's chamber of the pyramid:

> The Apostle Peter, reflecting on the importance of [the resurrection] exclaimed: ... whom God hath raised up, having loosed the pains of death: because it was not possible that he should be hidden of it. For David speaketh concerning him... Thou wilt not leave my soul in hell (*hades*, grave), neither wilt thou suffer thine Holy One to see corruption.
>
> Surely the coffer stands as a silent witness that Jesus indeed fulfilled the words spoken by King David, and quoted by Peter, for the gematria of the prophecy [Greek text] has the numerical value 8696. There are 8696 square MY [megalithic yards] in the surface area of the interior of the coffer.
>
> [2, pp. 88–89]

What would be the point?

Of course, humans cannot pretend to understand the workings of a divine intelligence, but I think numbers have clouded Ms. G.'s human intelligence. Numbers have the power to do that, too.

References

1. Gardner, Martin, *In the Name of Science*, G. P. Putnam's Sons, New York, 1952, reprinted by Dover, New York, 1957.
2. Gaunt, Bonnie, *The Magnificent Numbers of the Great Pyramid and Stonehenge*, Pyramid Research, Jackson, Michigan, 1985.
3. Kancil, Sang, in the *Malaysian Sun*, Wednesday, July 28, 1993.

CHAPTER 26
Pyramidiocy

Numerology, pyramidology, 666ology, and similar sciences are, thank heavens, on the fringes of intellectual life. We can hope that they will remain there, or, better yet, fall off the edge and disappear. Unfortunately, there are signs that their movement is in the opposite direction, toward the center.

When I first investigated numerology thirty years ago, I looked in the subject index of *Books in Print*, under "numerology," and bought all of the titles listed there. The expense was large, but it was not overwhelming since there were not all that many books listed. Those that were listed tended to be small volumes, cheaply produced on low-quality paper, sold by obscure publishers who mostly specialized in the occult.

Today I could not corner the market on numerology books, even if I wanted to. There are hundreds of them. They have titles like

The Numerology Kit
Numerology Has Your Number
Helping Yourself with Numerology
Numbers and You

Those are not put out by unknown publishers. The issuers of those four are, respectively, the New American Library, Simon and Schuster, Prentice-Hall, and Ballantine. Numerology is moving into the mainstream.

There are plenty of other books:

Win with Numerology
Numerology Asks: Who Do You Think You Are, Anyway?
The Sacred Science of Numbers
Racing Numerology: A System for Finding Winners of Horse Races Based on the Science of Numbers

Buying all of them would take many dollars.

Another example of the mainstreaming of nonsense is *Secrets of the Great Pyramid* by Peter Tompkins [2]. It was first published by Harper and Row in 1971, the paperback edition came out in 1978 and it is still going strong—in 1992 it was in its seventh printing. It is 420 pages long and beautifully produced with many illustrations. Its paper will not turn yellow for years, if ever.

But its contents are pyramidiocy, a descriptive term I learned from Martin Gardner.

> Recent studies of ancient Egyptian hieroglyphics and the cuneiform mathematical tablets of the Babylonians and Sumerians have established that an advanced science did flourish in the Middle East at least three thousand years before Christ, and that Pythagoras, Eratosthenes, Hipparchus and other Greeks reputed to have originated mathematics on this planet merely picked up fragments of an ancient science evolved by remote and unknown predecessors.
>
> The Great Pyramid . . . was designed on the basis of a hermetic geometry known only to a restricted group of initiates, mere traces of which percolated to the Classical and Alexandrian Greeks. . . .
>
> Like Stonehenge and other megalithic calendars, the Pyramid has been shown to be an almanac by means of which the length of the year including its awkward .2422 fraction of a day could be measured as accurately as with a modern telescope. It has been shown to be a theodolite, or instrument for the surveyor, of great precision and simplicity, virtually indestructible.
>
> [2, pp. xiii–xiv]

Though indestructible, it would not be easy for surveyors to use because of the difficulty in transporting it to the site to be surveyed.

> It is still a compass so finely oriented that modern compasses are adjusted to it, not vice versa. [2, p. xiv]

Adjusting the pyramid would be similarly difficult.

> It has also been established that the Great Pyramid is a carefully located geodetic marker, or fixed landmark, on which the geography of the ancient world was brilliantly constructed; that it served as a celestial observatory from which maps and tables of the stellar hemisphere could be accurately drawn; and that it incorporates in its sides and angles the means for creating a highly sophisticated map projection of the northern hemisphere. It is, in fact, a scale model of the hemisphere, correctly incorporating the geographical degrees of latitude and longitude. [2, p. xiv]

I suppose that I must say, though it should go without saying, that none of those things that Mr. T. (also the author of *To a Young Actress*, *Shaw and Molly Tompkins*, *The Eunuch & the Virgin*, *A Spy in Rome*, *The Murder of Admiral Darian*, and *Italy Betrayed*) asserts have been established are taken as true outside of occult circles. The world is filled with wonders, but not everything in the world is wonderful. The pyramid is wonderful enough without having silliness piled on top of it.

Even though Mr. T. knows hardly anything about what he is talking about, as will be extensively documented later, he is very sure of his ground:

> But certain facts must be confronted, and the textbooks amended to conform with them. Eratosthenes was obviously not the first to measure the circumference of the earth. Hipparchus was not the inventor of trigonometry. Pythagoras did not originate his famous theorem. Mercator did not invent his projection...
>
> Whoever built the Great Pyramid knew the dimensions of this planet as they were not to be known again till the seventeenth century of our era. They could measure the day, the year and the Great Year of the Precession. They knew how to compute latitude and longitude very accurately by means of obelisks and the transit of stars. They knew the varying lengths of a degree of latitude and longitude at different locations on the planet and could make excellent maps, projecting them with a minimum of distortion. They worked out a sophisticated system of measures based on the earth's rotation on its axis which produced the admirably earth-commensurate foot and cubit which they incorporated in the Pyramid.
>
> In mathematics they were advanced enough to have discovered the Fibonacci series, and the function of π and φ. What more they knew remains to be seen. But as more is discovered it may open the door to a whole new civilization of the past, and a much longer history of man than has heretofore been credited. [2, p. 285]

I am afraid that, as a result of the book's seven or more printings, many people now think, just because Mr. T. said so, that scientists are now all agreed that the pyramid is a theodolite, observatory, calendar, map, and geometry text, all in one large package. This is too bad. Since people's heads have a finite capacity, it is a shame to waste any of it with misinformation.

The trouble is, anyone can make assertions. For example,

> Funk-Hellet maintains that as early as the fourth millennium B.C. the Chaldeans had a mathematical series which gave the exact values of the cubit, the meter and π. What's more, he insists that the present *meter*, as developed by the French in the nineteenth century, was already a hermetic measure known in antiquity, and was linked trigonometrically with the cubit. [2, p. 264]

How pleasant if truth could be arrived at by maintaining and insisting! But it is not that easy.

Here is another assertion:

> Alvarez Lopez says that beginning with Mars, the planets are disposed in the order of the colors of the solar spectrum—with Mars as red, Jupiter yellow, Saturn yellowish-green, Uranus green, Neptune blue, and Pluto violet.

> According to Alvarez Lopez the Pyramid may once have been painted with the colors of the spectrum starting with red for Mars, just below the gilded pyramidion representative of the sun, and diminishing through yellow and blue to violet at the base, symbolizing the construction of the solar system both geometrically and with color. [2, p. 265]

How the pyramid-builders were able to ascertain the existence of Pluto, much less its color, is not explained.

After quoting the assertions, Mr. T. says

> Would it not be worthwhile, nonetheless, for academic institutions, so admirably equipped with computers and talent, carefully to analyze such conceits as those of Alvarez Lopez and Funk-Hellet and either refute them or support them with reliable data? [2, p. 265]

No, it would not. It *definitely* would not. Anyone may make assertions—an infinite number are available, many of them crack-brained—and life is too short to examine them all. Pyramidiots and other cranks do not understand this. They seem to think that the reason that no one takes the trouble to controvert their assertions is that they *cannot* be controverted.

Actually, they are not worth controverting. Here is the evidence for Mr. F.-H.'s assertion that the meter dates back thousands of years. It is that, to four decimal places, $\pi/6 = .5236$, and, he says, that the cubit used in the Temple at Jerusalem measures 523.6 millimeters. Therefore,

> A thousand years before Christ the Hebrews knew that a cubit was a mathematical entity dependent on the circumference, and were able to resolve π to four points of decimal. [2, p. 263]

The ancient Hebrews thus knew about millimeters. When the French National Assembly in 1791 defined the meter as one ten-millionth of the distance from the pole to the equator they must have suppressed the ancient Hebrew knowledge on which they were drawing. Of course, they did nothing of the kind. Mr. F.-H. has observed yet another instance of the Law of Small Numbers and that is all that needs to be said.

No one bothered to controvert another worker in the field of pyramidiocy, who bought advertising space in the *New York Times* to assert, as reported in [1], that the pyramids were built from the top down.

> The problem of how the pyramids were built has stumped archeologists, architects, engineers, mathematicians, and assorted savants for at least two hundred years. All of their laborious solutions have been grossly inoperable; mostly ramps and leverage devices that would require a million slaves a hundred years to build one pyramid. This problem has been publicized in numerous magazines, newspapers, journals and even on TV, and each new solution is debunked with little effort. Here is my solution judge for yourself!

> The pyramids were not built from the bottom UP, they were built from the top DOWN!
>
> They set a capstone on the peak of a suitable hill (probably amassed from eons of Nile silt). A "conveyor belt" of oxen brought the huge stones up the gentle slope and returned to the bottom with the dirt. The stones were placed around the capstone, whence the next level was dug, and so forth layer by layer, until the hill disappeared and the pyramid shone forth in its majesty. Less than two hundred stone masons, ox herders, differs, architects, supervisors, etc., would be needed. A child could do it. I won't insult you by pointing out the enormous advantages of this method, the tight logic of it, the easy answer to all objections, and how perfectly it lends itself to the interior details. Nobel prizes have been offered for less.

For the life of me, I cannot see how the stone in the middle of the bottom layer got there. It must have been pushed into place, but to do that would it not be necessary for something to have been holding up the rest of the pyramid?

The author could not get anyone to pay any attention to his assertions.

> Were I an archeologist, they would have called press conferences and proclaimed my idea around the world with fireworks. But I am not one of them. They answer with silence.

And quite rightly, too. That is what unsupported assertions deserve.

Mr. F.-H. sought support from numbers. He found some

> extraordinary results, including values for π, φ, and the royal cubit; this led him to conjecture that the royal cubit might have had a theoretical value before it had a practical one. [2, p.263]

He did this by making a Fibonacci-like sequence starting with 1 and 5:

$$1, 5, 6, 11, 17, 28, 45, 73, 118, 191, 309, 500, 809, 1309, \ldots$$

Notice that the product of three successive terms of the sequence, $11 \times 17 \times 28 = 5236$, the digits of the royal cubit. Notice that 45 is an eighth of a circle in degrees. Notice also, as neither Mr. F.-H. nor Mr. T. did, that these are manifestations of the Law of Small Numbers.

Now multiply the terms by two:

$$2, 10, 12, 22, 34, 56, 90, 146, 236, 382, 618, 1000, 1618, 2618, \ldots$$

There φ is! The last four terms are, to four significant digits, $1/\varphi$, 1, φ, and φ^2! Proof!

But there is even more, unsuspected by Mr. F.-H. Just before 618 comes 382, and that is the number of a course that I teach at DePauw University. What is more, the title of Mathematics 382 is... Elementary Number Theory! How did the ancient Hebrews know that?

Mr. F.-H. found even more. Multiply the terms of the original sequence by four and you get a series containing 5236, since $5236 = 1309 \times 4$. Multiply that by six and you have 31416, or π. All of this is the Law of Small Numbers in action, of course, combined with human ingenuity, as in starting the Fibonacci-like sequence with 1 and 5.

Besides applications of the Law of Small Numbers, Mr. T.'s book contains an enormous number of misstatements, misrepresentations, and downright errors. His unsuspecting readers, in the course of swallowing his conclusions, ingest them as well. If they knew about them, they might choke on what he concludes from them. But they will probably never know since when truth pursues error, it all too often happens that truth never catches up. Error has more wind.

David Singmaster, an English mathematician of great intelligence, wide knowledge, and energy (not to mention generosity, good humor, and diligence), took the trouble to go through Mr. T.'s book and, as lice can be picked in plenty from some people's heads, picked out some errors and corrected them. Notice that I did *not* say that Mr. T.'s book is lousy.

What follows are portions of the book followed by Dr. Singmaster's comments. I have no hope that they will undo much of the damage done by Mr. T., nor will they change his mind, but they serve as evidence of how careful we must be when reading books that make startling assertions.

> T: Till recently there was no proof that the inhabitants of Egypt of five thousand years ago were capable of the precise astronomical calculations and mathematical solutions required to locate, orient and build the pyramid where it stands. (p. xiii)
>
> S: Our factual knowledge about the mathematical abilities of the ancient Egyptians is primarily based on two papyruses: Rhind and Moscow, dating from about 1650 B.C. and 1850 B.C., but containing material possibly 300 years older. These reveal a fairly clumsy number system which could not have been used for astronomical calculations. There is no evidence of any theoretical geometry, though a value of $256/81$ was used for π. In particular, there is no evidence that the Egyptians knew any example of the theorem of Pythagoras, not even the 3, 4, 5, triangle.
>
> T: According to modern academicians the first rough use of π in Egypt was not until about 1700 B.C.—at least a millennium after the Pyramid; Pythagoras's theorem is attributed to the fifth century B.C.; and the development of trigonometry to Hipparchus in the second century before Christ. That is what the Egyptologists say, and that is what they put in their textbooks. (p. xiii)
>
> S: Pythagoras' theorem was well known to the Babylonians about 1700 B.C. The value of $256/81$, while an estimate of π, is not all that bad and indicates some previous thought and some method for finding it. The

Rhind papyrus, of about 1700 B.C., however does say that it was copied from material back to about 2000 B.C. and the relative sophistication of 256/81 would indicate that the cruder value of 3 must have been done earlier.

T: In 1671 Picard measured a degree of latitude between Amiens and Malvoisine.... On the basis of this computation Newton was able to announce his general theory of gravitation. (p. 32)

S: The new measurement of the earth did not cause Newton to publish his work. It was the famous encounter with Halley in 1684 and Halley's urging which got it published in 1685–7.

T: Not till the sixth century was π correctly worked out to the fourth decimal point by the Hindu sage Arya-Bhata. It took another thousand years before the Dutchman Pierre Metius calculated π to six decimals by means of the fraction 335/113. In 1593 François Viete carried the computation to eleven figures, and a generation later Rudolph Van Cuelin, just before he died, took π to 127 figures by postulating a circle with 36,893,488,147,419,103,232 sides. In 1813 the English mathematician William Shanks developed π to 707 decimals. Modern computers have carried the operation to 10,000 points of decimal, but with no solution to this apparently incommensurable number. (p. 71)

S: The Chinese had 3.14159 in 246 A.D. and 3.1415926+ in the 5th century A.D. Ludolph (not Rudolph) van Ceulen (not Ceulin) found π to only 35 places. The number of sides given by Tompkins is 2^{65}. My source says he used 2^{62} sides. Shanks published in 1873 (not 1813). He was wrong from the 527th place. π was computed to 100,000 places in 1962 and to 500,000 places in 1967 and probably to more places by now. π was shown to be irrational (i. e. incommensurable) by Lambert in 1767. What is a solution to a number?

[And, for that matter, what is "postulating" a circle with 2^{65} sides? Circles don't have sides, no matter how much you postulate. Mr. T. wrote a remarkable paragraph. *Everything* in it is wrong, with only a few exceptions. The grammar, for example, is fine.]

T: The coffer appeared to be designed to remain at a constant temperature and barometric pressure. (p. 83)

S: The notion of constant barometric pressure is absurd. Since the chamber is open to the outside, the barometric pressure is the same as that outside. The pressure could be kept constant only if the chamber were airtight.

T: But the cut was cunning in that it thus incorporated in the chamber both the 2-$\sqrt{5}$-3 and the 3-4-5 Pythagorean triangles. (pp. 101, 103)

S: There is no evidence that the Egyptians knew any example of the theorem of Pythagoras

T: Davidson claimed that—without getting into higher mathematics—it was evident that if you know the earth's distance from the sun and the length of the sidereal year in seconds, you can compute the rate at which the earth is falling into the sun. This in turn would lead to finding the specific gravity of the earth, of the sun, of the earth and moon combined, and even the speed of light. (p. 113)

S: Davidson's claims are mostly false. You can compute the centripetal acceleration of the earth toward the sun—"the rate at which the earth is falling into the sun"—but this is independent of the mass of the earth and of the sun, so the remaining items cannot be determined.

T: In the *Timaeus* he [Plato] explains the cosmos as being constructed by the triangle 3-4-5 and the number $\sqrt{5}-1$ or 1.236068 (which in common practice was taken as 1.2345). (p. 119)

S: It is impossible for the Greeks to have used 1.2345 for $\sqrt{5}-1$, since they did not have decimals. If anything, they would have used Babylonian sexagesimals: $\sqrt{5}-1 = 1.236068\ldots = 1 + 14/60 + 9/60^2 + \ldots$ with approximations of $1 + 14/60 = 1.233\ldots$ and $1 + 14/60 + 9/60^2 = 1.235833\ldots$.

T: Thom says the ancient engineers managed to raise perhaps ten thousand megaliths from one end of Britain to the other, and set them with an accuracy of 0.1. When they wanted to, says Thom, they could measure with an accuracy of 1 in 500. (p. 138)

S: An accuracy of $.1 = 1/10$ is not very good. But he can't mean .1% which would be 1/1000 since he says 1/500 was very good. I can't make sense of this.

T: Once the ancients had measured the length of the Descending Passage and its angle of descent, it would have been simple, by elementary trigonometry, to locate a central spot immediately above the end of the Descending Passage as a center for the proposed pyramid. (p. 151)

S: There is no evidence that they knew any systematic trigonometry

T: Once the diurnal pattern of the stars' apparent rotation past a fixed meridian had become clear to the observers, they could more easily plot the irregular and sometimes apparently retrogressive path of the planets and the moon in relation to the "fixed" stars. The heliocentric pattern of our solar system could well have been extrapolated from a study of the relative motions of these planetary satellites, anticipating Copernicus by 1000 years. (p. 156)

S: I doubt that the Egyptians could have deduced the Copernican theory. It is possible to make the Ptolemaic earth-centered theory as accurate as necessary—it's just a lot more complicated.

T: Stecchini found a glyph carved on the thrones of virtually all the Pharaohs since the Fourth Dynasty which contained geodetic data and hence astronomical data of extraordinary subtlety, enabling him to determine that the Egyptians used three figures for the tropic of Cancer: a simplified one of 24°, a precise one of 23° 51′, and one of 24° 06′ required for observing the sun's shadow at the summer solstice. (p. 177)

S: Since the Egyptians did not have degrees, it seems most unlikely that they would have used a figure rounded to an exact number of degrees, unless it were a simple fraction of a right angle. Their system of fractions would have expressed $24/90 = 4/15$ as $1/4 + 1/60$ or as $1/5 + 1/15$.

All of Chapter XIV is based on certain simple numbers of degrees and minutes. Since the Egyptians did not have these, the whole chapter seems rather improbable.

T: The slope angles at various heights also give important angles, such as $\sqrt{5} - 1$, which is also incorporated into the Great Pyramid. Such triangles, and the number $\sqrt{5} - 1$ (in common practice taken as the magic series 1-2-3) were fundamental in the operations of land surveying.

The third, fourth, and fifth steps of the ziggurat make triangles with sides related as the Pythagorean 3-4-5 triangle. (p. 187)

S: I cannot find $\sqrt{5} - 1$ or 3, 4, 5 anywhere in the figure. The Babylonians would never have used 1.23 for $\sqrt{5} - 1$, since they did not have decimals.

T: As the Babylonians liked to count by sixes, with a hexagesimal and sexagesimal system, the steps of the ziggurat rose in multiples of 6°. (p. 187)

S: The Babylonians did not count by 6's. They counted by 10's and 60's.

T: Thereafter the operation produced a simple progression: 8.666, 6.666, 4.666, 3.666, 2.666 for the cosine value of the angles indicated by each step.

The top step, says Stecchini, was rectangular instead of square, because the average of its sides gives 2.5833, which is the cosine of 75° 01′. (p. 188)

S: The cosine of an angle is always less than or equal to 1.000. $\cos 0° = 1.0000$, $\cos 33° = .8587$, $\cos 51° = .6293$, $\cos 57° = .5466$, $\cos 63° = .4540$, $\cos 69° = .3584$, $\cos 75°01' = .258538$. These values are not at all well explained by the "simple progression." The "average of the sides" giving 2.5833 is confusing. The only sides I can see are 4 and $2\frac{1}{2}$.

T: φ, like π, cannot be worked out arithmetically; but it can be easily obtained with nothing more than a compass and a straightedge. (p. 190)

S: The statement that φ and π cannot be found arithmetically is a confusion of the situation. Arithmetically, we can compute both to any number of places.

T: The odd, if not unique, mathematical fact that $\varphi + 1 = \varphi^2$ and that $1 + 1/\varphi = \varphi$ leads to an additive series, known as a Fibonacci series, in which each new number is the sum of the previous two: 1-2-3-5-8-13-21-34-55-89... etc., and their ratio comes closer and closer to φ. (p. 192)

S: The fact that $\varphi + 1 = \varphi^2$ is not at all odd as it arises directly from the defining relation on p. 191. It is unique since this equation has only two roots $(1 \pm \sqrt{5})/2$, and the negative one is not relevant here.

T: From the Arabs, Fibonacci learnt the system of numerals from 1 to 9, which he is credited with having introduced to Europe, where calculations were still being made by the clumsy means of Roman numerals and Greek letters. (p. 192)

S: Leonardo Pisano would have learned the numbers 0 to 9.

And Roman numerals and Greek letters were not used for calculations, but only for recording the results of calculations made using an abacus or similar device.

T: In the Renaissance the φ proportion, or Golden Section, as it was called by Leonardo da Vinci, served as the hermetic structure on which some of the great masterpieces were composed. (p. 193)

S: The term "Golden Section" did not come into use until the 19th century, some 300 years after Leonardo da Vinci.

T: Schwaller de Lubicz also found graphic evidence that the pharaonic Egyptians had worked out a direct relation between π and φ in that $\pi = \varphi^2 \times 6/5$. (p. 194)

S:$\pi = 3.141592654$, $6\varphi^2/5 = 3.141640787$, so we have an approximation accurate to 15 parts per million. There is nothing to connect the value of $6\varphi^2/5$ with π except the coincidental closeness of their numerical values.

T: *En passant*, Schwaller notes the curious coincidence that the Greeks should have adopted for the relation of diameter to circumference the symbol of π, which looks just like the Egyptian's doorway. (p. 195)

S: The Greeks did not adopt the symbol π. To them, it would have represented the number 80, because of their alphabetic number system. It was introduced in the 18th century A.D.!

T: Schwaller checked several score of these royal napkins for hermetic significance and found that they invariably gave two angles whose values were respectively φ and $\sqrt{\varphi}$. (p. 195)

S: He confuses the notions of angle and the trigonometric ratios of an angle.

T: Although the squaring of the circle is an insoluble problem if you use the irrational number of π, it nevertheless practically resolvable as a function of the Golden Number φ. (p. 197)

S: He clearly states that squaring the circle is impossible and that his descriptions are only practical. Later he tends to overlook this point.

T: With the Pyramid, the ancient Egyptians had not only squared the circle but effectively cubed the sphere. (p. 200)

S: He leaves out "practically." He does not actually ever find a volume, so his reference to cubing the sphere is not justified. He has shown only how to find the area of the surface. On p. 199 he refers to this as the "main problem of the map maker," but this is not even a minor problem—did you ever see a map the size of the earth?

T: Agatharchides reported that the length of one side of the base of the Pyramid corresponded to $1/8$ minute of degree, and the apothem to $1/10$ of a minute.

Jomard had hit upon this information and used it to find an almost exact solution: his apothem of 184.722 meters multiplied by 10 gave a minute of 1847.22 meters, which is almost precisely the length of a minute of latitude at the twenty-ninth parallel. (p. 201)

S: The lengths $1/8, 1/10$ give base/height $= 10/\sqrt{39} = 1.6013\ldots$, a value different than the four previous ones.

T: Translated into meters by multiplying by .308 the result is 39,916.8 kilometers, which is within one quarter of one percent of our modern earth circumference of 40,000 km. (p. 207)

S: 40,000 is a rough estimate for the circumference of the earth. On p. 264, he gives 6378 km as the radius which gives 40,074 as the circumference.

T: At the thirtieth parallel it would vary from the equator by the cosine value of 30° or $\sqrt{3/2}$ (p. 211).

S: He misprints $\frac{1}{2}\sqrt{3}$ as $\sqrt{3/2}$.

T: ... when Alexander the Great destroyed Persepolis in the fourth century B.C. he may have exterminated the Egyptian geographers imported by the Persians to do their figuring. (p. 214)

S: It is most unlikely that anyone would have used Egyptians to do figuring. The Egyptian number system was so useless for such calculations that it was replaced, even in Egypt, by Babylonian numbers when serious astronomical calculations began (about the first century A.D.). The Egyptian culture had held on to their numbers for some time after Alexander, much as it had held on to the Bronze Age for a thousand years after iron was widespread elsewhere.

T: [The passage cited earlier about the pyramid being painted the colors of the spectrum to match the colors of the planets]

S: Lopez' theory neglects the point that the last three planets could not have been known until there were good telescopes. Uranus was found in 1781.

T: Denis-Papin says that the Arc was placed in a dry spot where the magnetic field reached a normal 500 to 600 volts per vertical meter. (p. 278)

S: Magnetic fields are not measured in volts per meter—electric fields are.

The "Arc" is the biblical ark of the covenant. The revised spelling is probably intentional.

T: In mathematics they were advanced enough to have discovered the Fibonacci series, and the function of π and φ. (p. 285)

S: He has presented no evidence that the Egyptians knew the Fibonacci numbers. Knowledge of φ does not involve knowledge of the Fibonacci numbers—witness the Greeks, who were keen on it but did not know of the Fibonacci numbers.

T: But in terms of latitude the apex appeared at first not as perfect as it should have been, since it was at latitude 30° 06′ north and not at the perfect latitude 30° 00′ north, which is the latitude of the Great Pyramid of Giza. But the Egyptians reassured themselves by observing that the southern limit of Egypt is indicated by the First Cataract. The upper edge of this cataract is at the perfect latitude 24° 00′ north, whereas the lower edge is at 24° 06′ north. Hence they could say that Southern Egypt has an extension of 6° 00′,... (p. 293)

S: There is no evidence that the Egyptians ever used degrees, so most of [this] appendix is based on false usage of simple numbers of degrees.

S: As a brief conclusion, let me say that there is no evidence for the amazing mathematical and astronomical abilities of the ancient Egyptians, which are used in this book to "explain" the great pyramid. None of the material which has survived indicates even an interest in such observations and calculations. The author has done a lot of research into past work, but his presentation shows little understanding of school algebra and trigonometry and little reading in the history of mathematics. Consequently most of his arguments are little more than wishful thinking combined with coincidental similarities of numbers.

Nevertheless, the pyramid remains enigmatic and shows that the Egyptians were masters of practical engineering and organization. I am even willing to grant that the base/height ratio may have been designed to be $\pi/2$ as they knew it. But I cannot see that they knew any astronomical data, beyond the length of the year, of the sort attributed to them in this book.

There are other errors that Dr. Singmaster missed, such as the φ-foolishness that Le Corbusier used the golden ratio in designing the United Nations Building in New York, but pointing them out is not necessary. The list of errors is long enough already to show that Mr. T. is not to be depended on in matters of geometry, mathematics, and astronomy. Nor, as far as I am concerned, on pyramids either.

His defense, if he went so far as to admit that he made errors, would no doubt be that such things are mere trivialities, minor errors that do not detract from the force and importance of the book's main message. Maybe so, but I think not. The trouble is that truth usually cannot catch up with error and, even when it does, error can wiggle away and escape. Nevertheless, we must keep up the pursuit, as relentlessly as we can, as long as we can.

References

1. Carroll, Jon, Pyramid power; amazing revelations, San Francisco *Chronicle*, December 2, 1992.
2. Tompkins, Peter, *Secrets of the Great Pyramid*, Harper and Row, New York, 1971; paperbound edition 1978.

CHAPTER 27

Are You Gridding?

Most purveyors of numerology would have us believe that they are passing on ancient and arcane wisdom handed down to them, if not directly from Pythagoras, then at least through a chain of intermediaries wise in the ways of the occult. We know, even when they do not, that it was Mrs. L. Dow Balliett of Atlantic City who was the source and not Pythagoras, except very indirectly. But it is more impressive to cite Pythagoras and other high authorities, all the higher for their being two or more millennia in the past and not from New Jersey.

However, though they have many points of similarity, numerologists are not all alike. Some make no claim to reveal ancient wisdom and plunge right in to their subject, making assertion after assertion, never mentioning where their information comes from. This makes odd reading to those of us trained in mathematics, told as we constantly were to accept nothing on faith or authority and to prove all things. (Even when the author of the mathematics book we are studying tells us that a proof is beyond the scope of the text we are given places where we can find the proof, though the author knows, and we know, that we would not understand them if we did look them up.)

When reading numerology books, "Why?" keeps springing to my mind, along with "How come?" and "How do you know that?" Evidently such questions do not bother those who buy and read books by numerologists. For them it seems that assertion suffices to guarantee truth. Perhaps they are practical people, concerned only with results. Theory can be so dull.

Austin Coates, the author of *Numerology* [1], does indeed plunge right in—the first sentence of chapter 1 is,

> In order to understand the meaning of the nine numerals in relation to human personality, they have to be set out as a grid, thus:
>
> 3 6 9
> 2 5 8
> 1 4 7

There is no indication of *why* they "have" to be set out that way. They just do. No other writer on numerology before Mr. C. put the digits into a matrix, perhaps because other writers on numerology all copy from each other.

But Mr. C. is original. His publisher says,

> His system has evolved over a period of forty years of "doing people's numbers" as a pastime, and is based solely on personal experience, on observation of the behavior of numbers in relation to human personality. At the time of writing he had not read any work on the subject, nor had he ever met or communicated with another numerologist.

Writers of mathematics texts, even those with forty years' experience with their subject, never brag about their ignorance of what they offer instruction in. This is one more difference between books on mathematics and on numerology, and between mathematicians and numerologists.

Based on his observation of the behavior of numbers, Mr. C. asserts that the numbers in the three horizontal and three vertical lines have characteristics.

> The line 3, 2, 1 consists of numbers of thought, connected with the element of air. The line 6, 5, 4 consists of numbers of activity, connected with the element of earth. The line 9, 8, 7 consists of numbers of power, connected with the heaviest of the elements, water.
>
> Needless to say, there is no actual connection between numbers and elements. It is simply that associating the two provides a convenient method of explaining certain aspects of numbers. [1, p. 3]

Needless to say, there are no reasons given for the assignments, so they must have been arrived at mystically. But after they are made, we have a diagram [1, p. 4] that, Mr. C. says,

> is the key to the meaning of numbers, and is the basis of numerology

	Thought *Air*	*Activity* *Earth*	*Power* *Water*
Sunlight *Head*	3	6	9
Produce *Heart*	2	5	8
Soil *Stomach*	1	4	7

This is new, this has not been seen before. The mind-spirit-body triad (head-heart-stomach) is of course old, but sunlight-produce-soil was a new one for me, as was the idea that as sunlight and soil combine to produce produce, so then must head and stomach combine to produce heart. Of the four elements—earth, air, fire, and water—one had to go, since there is no way to make four elements and nine digits fit symmetrically together. Fire was left out though it seems to correspond better to activity than earth does.

Mr. C.'s new system works just as well as those of more standard numerologists. This implies that it is not the ancient wisdom and spirit of Pythagoras that makes numerology work, but something else. The something else, of course, is the human ability to see things that are not there, the human desire for wonders and marvels, the human tendency to believe what authorities say, the human preference for the ease of magic over the discomfort of reason. The human race has not yet evolved far enough so that reasoning and rationality are natural. We may use them, but we do not like them. They are hard, they are work, their rewards can seem poor and meager; it is too much to ask us to stay within their bounds. We *must* break out. Even some mathematicians have been known to be irrational when away from their mathematics.

Although Mr. C. is a numerologist, and a new kind of one at that, he shows that numerologists are not necessarily people with wild eyes and funny clothes who live in California. His father was Eric Coates, a composer of some distinction. He spent his adult life in Asia:

> He was Assistant Colonial Secretary and a Magistrate in Hong Kong for seven years and then Chinese Affairs Officer and Magistrate in Sarawak. In 1959 he became First Secretary to the British High Commission in Malaya.

He writes with clarity and grace, as befits the author of *Invitation to an Eastern Feast*, *Personal and Oriental*, *Basutoland*, *Prelude to Hongkong*, *Western Pacific Islands*, and many other books. If you encountered him in person, I am sure that you would never guess that he was a numerologist.

But he is, and he is the originator of the grid system of numerology. Once you have the grid, here is what you do with it. Take each letter in a name and assign it a number, using the usual scheme modulo 9,

1	2	3	4	5	6	7	8	9
A	B	C	D	E	F	G	H	I
J	K	L	M	N	O	P	Q	R
S	T	U	V	W	X	Y	Z	

This is the system used by Mrs. L. Dow Balliett and almost every modern numerologist since, but it is sufficiently natural that it could have occurred to Mr. C. independently. But, being a deeper and better thinker than other numerologists, Mr. C. considered some questions that other numerologists often ignore. One, which no other numerologist has bothered with as far as I know, is what do you do with, say, the name of a Spaniard, whose alphabet has fewer than 26 letters? The answer is, use the 26-letter alphabet, because

> in these languages, persons of Germanic, Hungarian, or English ancestry have names which include the missing letters, thus these are effectively in use—as in telephone directories, for instance. [1, p. 35–36]

The question of what to do with letters like "ñ," "ü," and "é" he does not consider. The reason for this is that numerologists are almost all concerned only with how names look and not how they sound (chapter 23 has the only exception that I know of), so "ñüé" would have the same value as "nue," $5 + 3 + 5$.

Another question is, what do you do with the name of someone who does not write using the Roman alphabet?

> In Asia, use the full Roman alphabet for Overseas Chinese and others who normally sign their name in English, French or Dutch spelling and who *think* of their name in Roman script. [1, p. 36]

A third is, what name should you use? The general run of numerologists use the full name as it appears on the birth certificate, giving no explanation. This follows the practice of the founder, Mrs. Balliett, but she explained it by saying, in effect, that the Name is the Name, and the weightily sacred and mystical act of Naming is done once and for all. There is something to be said for that view—you do not get to choose your astrological sign, nor change it if you want a new one—but I think that there is more to be said for Mr. C.'s view that your numerological name is the name that you think of yourself as having.

> Numerology is really a numerical analysis of how we see ourselves. When asked to state our date of birth, which of the world's various calendars do we use? Which numbers do we instantly *see* as being associated with us? When asked to state our name, how many of our given names do we state? How do we *see* ourselves, where our name is concerned? Do we think of ourselves as Reginald Arthur Friedlander Pope? Or do we see ourselves as Reginald Pope?
>
> The commonest first question a numerologist is asked is, "What name should I give?" The answer to the question is, "How do you *see* yourself? By what name do you *think* of yourself?"
>
> Because numerology is no more than an analysis of how we see ourselves in relation to numbers and sounds (names), it is of major importance to get this question answered satisfactorily.
>
> With people who are not in the public eye—i.e., most people—this can be surprisingly difficult, especially in countries where it is the practice for children to receive many given names. There is also the difficulty that when asked, "What name do you *think* of yourself as?" an amazingly high percentage of people have *never* thought.
>
> Faced with this, there are various questions one can ask:
>
> "How do you sign a cheque?"
>
> "What name is on your passport?"

> "If a police officer suddenly asks you to state your full name, how do you reply?"
>
> "In a dream, if someone in authority calls you by name, what name does he use?"
>
> This last, if the person is capable of answering it, is probably the best. The essential is to reach the person *as he sees himself.* [1, p. 31–32]

On the other hand, your name is not completely under your control. Mr. C. later writes, also in contrast to other numerologists,

> Never change your name, in the hope of obtaining more favorable numbers. If you are a woman, your numbers will change anyway on marriage, and this is difficult enough*. Do not make it more difficult than it need be.
>
> Numerology is simply arithmetic in another guise. Feed it into a computer, and with correct programming, the machine would instantly issue accurate character readings.† It would also, with similar accuracy, predict—as we do when looking at statistics. Your numbers *are* your statistics.
>
> Deliberately change your name, and the computer would issue the character of a slightly different person. But you remain *you.* All you have done, therefore, is to give yourself a name which is not quite *you*, which is a nuisance and pointless.
>
> *The change of numbers on marriage simply reflects an actuality. You leave one house and family, and enter another. Numbers convey implacable truth.
>
> †I would not advise trying this, however. I suspect the readings would come forth as a series of maledictions, the machine not having the numerologist there to soften the blow. Grids contain a great deal, far more than is in these pages. When doing a person's numbers, one tells them only certain selected things, giving encouragement or warning. The rest is best left unsaid. Numerals have no conscience. [1, p. 91]

Though they may lack a moral sense, numbers do not lack meaning. For example, referring to the grid, we see

> In SIX the sunlit head is in conjunction with earthy activity. The head, full of ideas, here applies itself to the activities of the earth. This is the number of commerce, enterprise and law.
>
> In SEVEN the earthy stomach lies conjoined with watery power, a situation liable to end in mud. This is the number of agriculture: human power applied to the soil. [1, p. 5]

And so on: to be able to make such interpretations requires long meditation on the mystical associations of the digits and the skill cannot be learned quickly. This is one reason why there are no child prodigies in numerology, unlike chess, or mathematics. The properties of the digits, properly interpreted, give the grid, Mr. C. says [1, p. 7], "an additional set of dimensions" as in Table 1.

	Individualists	*Organizers (of things)*	*Influencers*
Independents	3	6	9
Organizers (of people)	2	5	8
Reformers	1	4	7

TABLE 1 Interpreted Grid Characteristics.

Once you have all that straight in your mind, what you do with your name, once you have *that* straight in your mind, is assign the proper number to each letter and then put circles around the numbers in the grid, one circle for each letter. For numbers with more than one letter, the circles are concentric. For example, the grid for Sir Christopher Wren is gotten from

C	H	R	I	S	T	O	P	H	E	R	W	R	E	N
3	8	9	9	1	2	6	7	8	5	9	5	9	5	5

and looks like Figure 1. The diagonal line is the *stress line*, drawn through the three numbers—horizontal, vertical, or on the two main diagonals—that have the maximum number of circles. If there is a tie for the maximum, there can be more than one stress line. Actually, it is a directed line, going from the weightier end, for Sir Christopher the 9, toward the lighter one.

Most of Mr. C.'s book, as with most numerology books, is devoted to explaining the significance of various possibilities. He gives interpretations for each of the nine digits, separately for males and females, and then for the eight stress lines (or sixteen, if you count both directions).

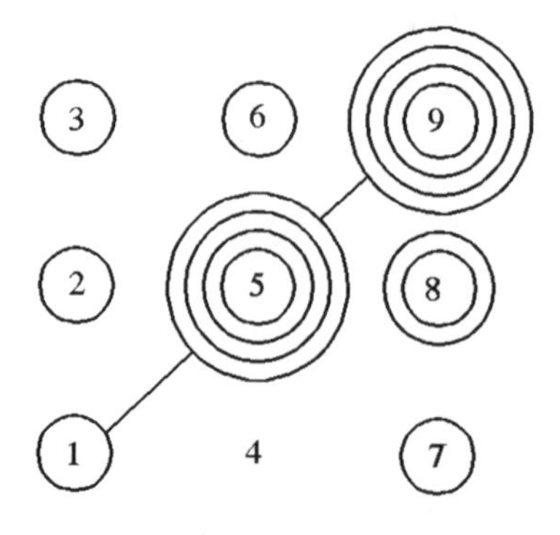

FIGURE 1 Christopher Wren.

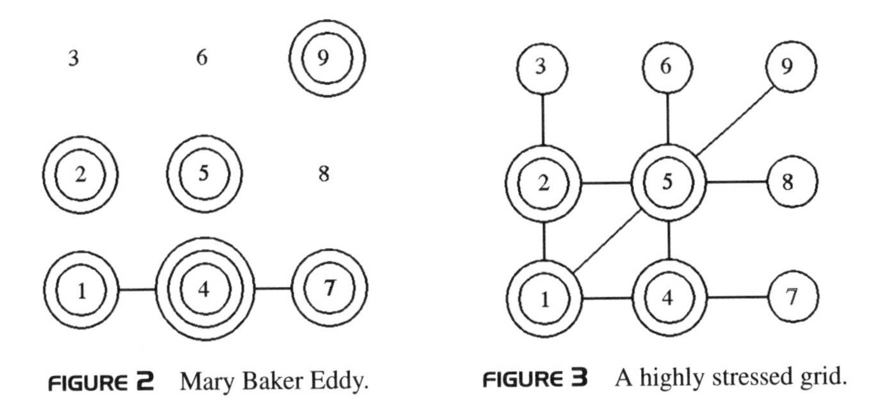

FIGURE 2 Mary Baker Eddy.

FIGURE 3 A highly stressed grid.

It is likely that Mr. C. forgot about the time he spent as a schoolboy evaluating determinants of three-by-three matrices. If he hadn't, he then could have completed his system by adding interpretations of the broken-diagonal stress lines. For example, the grid for Mary Baker Eddy in Figure 2 would have as an additional stress line the 2-4-9 broken diagonal. It does not seem fair that 5 should be on *every* diagonal stress line. But, being in the center of the digits, perhaps it deserves a central place.

The unfortunate person whose grid is in Figure 3 has *five* stress lines which, as Mr. C. says [1, p. 90], "is another instance of a grid containing too much."

Mr. C.'s reading of Wren's grid is as follows:

> In this, the weight is in 9 (arrow diagonally downward). This indicates a person usually born in favorable circumstances (9, power and money), capable of putting across ideas which he may not have invented, and which may not necessarily be new, but who does so with effect and concentration. He is a person prepared to make material sacrifices for what he believes in, and is inclined to do so in disregard of his wife and family.
>
> Equally, from a sound background (9), he is a person prepared to knuckle under to adverse conditions, and will expect his family to do likewise....
>
> Note the massy weight of 9 (power) and strength of expression (5), also the perfect balance on 3, 2, and 1, indicating great mastery in the arts and sciences. This stress line, weighted in 9, indicates integrity and absolute need to make up one's mind without interference.
>
> With everyone in this line, whether weighted upward or downward, a guiding urge is the exposition of truth. Whereas, weighted in 1, it leads from the particular to the general, when weighted in 9, it leads from the general to the particular.

> In Wren's case, it was extreme concentration on projection of style, to a degree at which, by looking up at a ceiling, one could see from its proportions that only one man in the world could have designed it.
>
> And this was achieved despite having to knuckle under throughout his life to the inferior opinions of men who did not share his vision. St. Paul's is Wren's splendid monument, but had he been allowed to build it in his own way—the plans, known as the Great Model are extant—it would have ranked as one of the most perfect structures on earth.
>
> His absence of 4 means impatience with detail, and in this position as the only missing number it usually implies good health. An intriguing aspect of it here is that Wren gave up astronomy (4) in favor of architecture.
>
> [1, pp. 45–46]

Seemingly an impressive reading. But is it really? It certainly is in breadth of knowledge and ease of expression when compared with other numerologists, but I think that it is at bottom like all the others. That is, it is full of statements that are easy to apply to many people and easy for many people to agree with. If you did not know that Sir Christopher Wren was the person being described, could you not have taken it as a description of yourself? I know that it fits me in many, many particulars, especially those that refer to virtues. It even fits some that are not necessarily virtues, such as impatience with detail. Impatience with detail—does not that describe you as well? It could very well, since *detail* carries with it the connotation of fussy, empty bother and we are all impatient with that. Finding good health in a grid is also a safe bet since people feel well, on the average, 95% of the time.

Mr. C. added a footnote to the Wren reading:

> The President of the United States, Richard M. Nixon, has this line, and the members of his family have experienced precisely its effects.

Mr. C. wrote that in 1974, while Nixon was still President. Let us see if it stands up. For everyone with the 9-5-1 stress line, Mr. C. said, a guiding urge is the exposition of truth. And, he said, it indicates integrity. Does this describe the Richard M. Nixon of Watergate and its aftermath? Not to me, it doesn't. You might try to make it do so by saying that, while Nixon had the *urge* to tell the truth, at times other urges, such as self-preservation, were stronger. Or, you might say that he was indeed telling the truth at all times, as *he* saw it. But it would be better, I think, if you concluded as I did that, stress line or no, Mr. C. has given a flat-out wrong reading of Nixon.

Despite your opinion of President Nixon, or of mine, the point is that Nixon would not think that Mr. C. was wrong. He would most likely have agreed enthusiastically that he had integrity and was always guided to expose the truth. He might even have admitted that the ideas that he put across with effect and concentration were not necessarily new. I am sure—I heard the Checkers speech—that he would agree that he had been made to knuckle under to adversity many, many times.

I think that Nixon would have said "Yes! That's me! How could you know all that?" I think that Christopher Wren could well have done the same, though he would more likely have said "That is I!" I think so because of the heights, or depths, to which human self-delusion can go. They are unscalable, or unfathomable, and they are part of what keeps numerology books selling briskly. (I have two copies of Mr. C.'s book, one a fifth printing and another a 1991 paperback edition. It has had British editions as well.)

Another reason for Mr. C.'s successes with numerology—you don't pursue a hobby for forty years if it provides only failure—comes from something that Mr. C. tosses in as an aside:

> If you wish to give sound advice by means of numerology, it is preferable to meet the person first, or in the case of an infant, to meet the parents, in order to know from what position in society one is starting. [1, p. 90]

Of *course* it is. It is even better to give the reading with the person present, so that you can see how what you say is taken and then quickly back off from anything that evokes disagreement, whether expressed in words, in an expression, or in a slight shifting of the body. Similarly, if you see that you are hitting the mark, you can repeat, emphasize, and expand.

Even so, you may sometimes miss. But, if you are clever, you never *really* miss.

> As an example, one may have a man who is a 6 by name and whose general numbers suggest the law or commerce, yet who is in fact an actor. Inquire closely, however, and it may be found that the actor is regularly engaged to play the parts of lawyers and businessmen, because this is the type of personality he conveys. Equally, one may find a person with the numbers of a wise statesman, yet who is in fact a village policeman. Yet inquire deeper and it may be found that he has a statesmanlike approach to the problems of his village, and is in fact exercising this ability in his own way.
> [1, pp. 123–124]

I am sure that each of us can look within ourselves and see those things that the numbers say, as interpreted by Mr. C., even if they are visible to no one else. Certainly I could have been a wise statesman—could not you as well? Even the shyest and most diffident can see themselves on the stage. The law? Certainly: I have an instinct for justice and a keen and logical mind. I bet that you do too. Architecture? You should have seen the constructions I made with my blocks as a child.

It is in a way too bad that this book is such a pleasure to read. Talent should be used for better ends. In the typical numerology book you do not learn that with *Cavalcade* (1931), Noel Coward

> ceased to be the brittle young man, and became a national figure.
> [1, p. 107]

Or that, still referring to Coward,

> were it possible to teach playwriting—which it isn't—*Tonight at 8:30* would be a textbook. As an example, *Hands Across the Sea*: how to write a play about absolutely nothing, and keep an audience shaking with laughter from start to finish. [1, p. 108]

No other numerology author that I know of would be capable of writing,

> My aim here is to show a man's life, as he himself might want it to be.
>
> He does not wish to be assassinated (John F. Kennedy), or end his life in the remote and awful grandeur of St. Helena (Napoléon). He will wish to die peacefully, in his own home, surrounded by love, and with the sense of having accomplished something. [1, p. 95]

References

1. Coates, Austin, *Numerology*, Citadel Press, Secaucus, New Jersey, 1974.

CHAPTER 28

Enneagrams

Hippocrates divided people into four categories by temperament (sanguine, phlegmatic, choleric, and melancholy) based on which of their bodily fluids (blood, phlegm, yellow bile, and black bile) was dominant. There were four fluids because there were four elements (earth, air, fire, and water) and the ancients had some nice correspondences:

earth	black bile	melancholy	cold
air	yellow bile	choleric	dry
fire	blood	sanguine	hot
water	phlegm	phlegmatic	wet

Even though we now have more than four elements, Hippocrates' types still have representatives among us and, if you like classifying, can still be used for classification. The gain in doing that is more or less the same as the gain in giving something a name. It does not add anything to understanding, but giving things names may give us the illusion of having more control over them.

Followers of the enneagram movement increase Hippocrates' number of categories from four to nine and, unlike Hippocrates, give them numbers.

Here they are, from [2]:

1. The Perfectionist. Conscientious, rational, critical and rigid.
2. The Giver. Empathetic, demonstrative; can be intrusive and manipulative.
3. The Performer. Competitive, efficient, Type A, obsessed with image.
4. The Romantic. Creative, melancholic, attracted to the unavailable.
5. The Observer. Emotionally remote, detached from people and feelings; private, wise.
6. The Questioner. Plagued by doubt, loyal, fearful, always watching for signs of danger.
7. The Epicure. Sensual, cheery, reluctant to commit.
8. The Boss. Authoritarian, combative, protective, take-charge, loves a good fight.

9. The Mediator. Patient, stable, comforting; may tune out reality with alcohol, food or TV.

How the discoverers of the enneagram decided to assign numbers to the nine characters I do not know. They have parallels with the numerological properties of the digits (see chapter 20 for the nearest summary): 2 and 8 are perfect matches and 1 and 7 come close. There are also differences—the numerological 5 is nothing like the enneagramatical 5.

With hardly any straining, most people could fit themselves into many of the types. I find that I am a 1, 5, 9, 4, 8, 3, 7, 2, and 6, though not necessarily in that order, but only for the positive qualities.

Enneagrams are having a vogue at the moment, making the pages of *Newsweek*, from which the categories above were copied. There are more than thirty books on the topic, which have sold more than one million copies. There are enneagram workshops. *The Enneagram: A Journey of Self Discovery* [3] has been translated into Croatian, French, German, Italian, Japanese, and Spanish. The First International Enneagram Conference, held in California in August 1994, attracted 1400 participants.

The attraction of the enneagram, and what people get out of it, seems to be self-knowledge. The question whether the self-knowledge is real or illusory seems to arise as much for the enneagram as it does for numerology or astrology. But enthusiasts of the enneagram are sure that they have benefitted. One is quoted by *Newsweek* as saying "Sometimes I catch myself being too demanding in my work or my marriage, but the Enneagram opens the door to compassion." Beyond self-knowledge are applications: a partner in an executive search firm finds the enneagram "a powerful hiring tool." How the tool was applied was not specified.

You cannot produce thirty books, selling a million copies, merely by saying that people may be divided into nine personality types any more than numerologists can with their nine types. You have to go into relations, connections, and so on. If you draw a regular nonagon and put the numbers from 1 to 9 on the vertices, then you can connect them as in Figure 1, the fundamental enneagram diagram.

For all that most enneagramists know, the diagram could have been found carved on a stone tablet somewhere. This is not the case. Its originator was a Russian-Armenian, George Ivanovich Gurdjieff (1866–1949), a successful cult leader who I think had many similarities to Pythagoras. He wanted people to think that the enneagram came from the ancient wisdom of the East, but there is no evidence that it is old. Hippocrates attached no integers to his four temperaments. Gurdjieff may have made it up, but whether he made it up or, as he claimed, found it in India, he is to blame for making it popular.

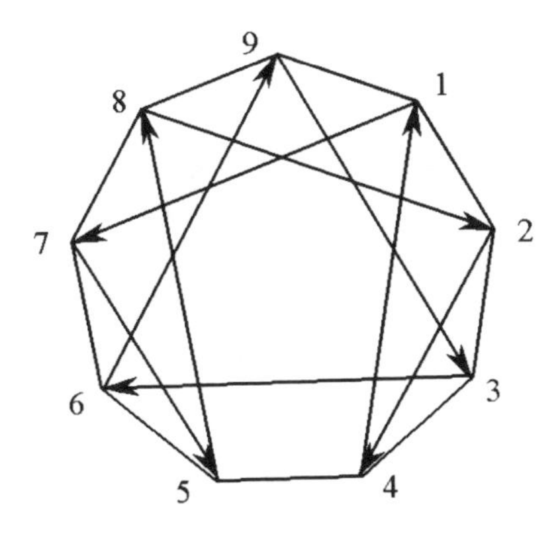

FIGURE 1 Enneagram connections.

Gurdjieff founded a movement that promised mystical enlightenment and he attracted, as cult leaders with powerful personalities can, numerous followers. Gurdjieff's included some prominent people, including Frank Lloyd Wright, Georgia O'Keeffe, and Katherine Mansfield, who unfortunately died while undergoing the rigors of residence at his Institute for The Harmonious Development of Man in France. His ideas, promoted and popularized by one of his disciples, the journalist P. D. Ouspensky (1878–1947), live on in the Fourth Way movement, whose doctrines have no numerical content.

Here is a picture of Gurdjieff in action [5, p. 96], quoted in [1, p. 14]:

> "You are a turkey cock," he said to someone the first evening. "A turkey cock pretending to be a real peacock." A few masterly movements of G's head, a guttural sound or two, and there appeared at the table an arrogant gobbler parading itself before a hen. A little later a much larger animal materialized itself before our eyes. "Why do you look at me as one kind of bull looks at another kind of bull?" he asked of someone else. And with a slight change in the expression of his eyes, in the carriage of his head, and in the curve of his mouth a challenging bull was produced for our inspection.

Pythagoras probably caused similar effects among his followers.

Enneagramism is free of numerology in not determining your category by the number of your name, or your date of birth, but it contains number mysticism nevertheless. The pleasingly symmetric arrangement of the lines in Figure 1 was not made arbitrarily. As one enneagram book puts it,

> the nine-pointed star maps the relationship between two primary laws of mysticism, the law of Three (trinity), which identify the three forces that are present when an event begins, and the law of Seven (octaves), which governs the stages of implementation of that even as it is played out in the physical world....

> The central triangle of Three-Six-Nine can also be mathematically described as the attempt of the trinity of forces (elsewhere explained as Father, Son, and Holy Ghost; Brahma, Vishnu, and Siva; creative, destructive, and preserving; active, receptive, and reconciling; or (as Gurdjieff called them) force One, force Two, and force Three)) present in the original creation to be reconciled back into one. This is illustrated arithmetically by dividing 1, or unity, by 3, which results in a fraction, the last numeral of which repeats infinitely, that is, $1 \div 3 = 0.3333\ldots$.
>
> Once an event is begun, the law of Seven, or law of octaves, comes into play. The law of octaves is preserved in the musical scale as Seven notes with a repeating Do and governs the succession of stages by which an event is played out in the material world. The relationship of Seven to unity can be expressed by dividing 1 by 7, which yields the repeating series 0.142857142857... which contains no multiples of three. The full Enneagram is a circle divided into nine equal parts that represents the fusion of the law of Three and the law of Seven, which interact in specific ways along the diagram's inner lines. [1, pp. 36–37]

That is why the lines go through vertices 1, 4, 2, 8, 5, 7 and why the arrows point in the directions that they do. Personalities are related the way that they are *because* of the numbers in the decimal expansion of $1/7$. This is numerology. The numbers take over. It is, I suppose, conceivable that the personality types were assigned their numbers just so that the arrows determined by 1/7 would make sense, mystical or otherwise. If that was so, the numerology is borderline and not pure, but I suspect that the numbers bent the connections to their will. Numbers, as I may have said, have power.

The enneagram is the result of a physiological happenstance. If we had six fingers on each hand instead of five, we would count by dozens instead of tens and the mystic polygon would have eleven sides instead of nine. If we counted by twelves, our digits would be

$$1, 2, 3, 4, 5, 6, 7, 8, 9, \mathcal{X}, \text{ and } \epsilon.$$

The last two are pronounced "dec" and "el". The digits have no division quite as slick as that of the base-10 digits into thirds (.333..., .666..., .999... for 3, 6, and 9) and sevenths (.142857... for the others). The unit fractions in base 12 are

n	2	3	4	5	6	7	8	9	$\mathcal{X}$	ϵ
$1/n$	.6	.4	.3	$.\overline{2497}$	.2	$.\overline{186\mathcal{X}35}$	.16	.14	$.\overline{12497}$	$.\overline{1}$

It looks as if the best that could be done is to use $1/7$ and $1/5$ (the Law of Seven and the Law of Five) to divide the digits into three classes, $\{1, 8, 6, \mathcal{X}, 3, 5\}$, $\{2, 4, 9, 7\}$, and poor lonely $\{\epsilon\}$ leading to Figure 2. As Figure 1, it has some nice symmetries and pleasing parallelisms, illustrating the power of numbers

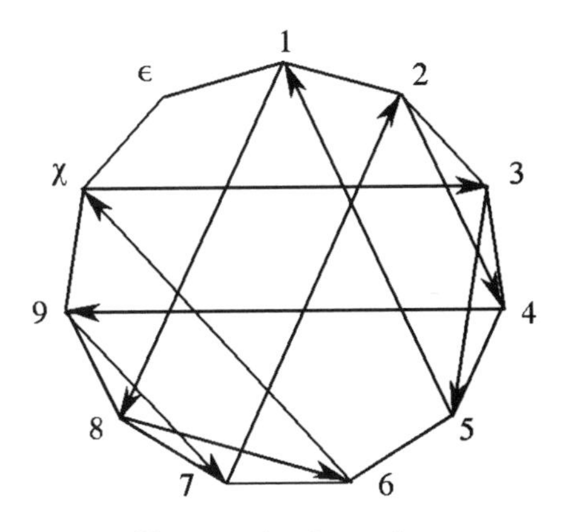

FIGURE 2 A twelve-fingered enneagram.

to do the unexpected: I had no idea, when choosing my partition of base-12 digits, that it would lead to such a pretty picture, one that might yield mystical insights if contemplated long enough. Category ϵ is obvious: The Loner.

Enneagram doctrine holds that

> when relaxed, you take on positive traits of the personality type your arrow points toward. Under stress, you assume negative traits of the type whose arrow points toward you.

This is illustrated in Figure 3.

The books on the enneagram that I have seen do not go into any of the theoretical bases of the system. In common with astrology and numerology, their assertions were simply that—assertions, whose verification I suppose was

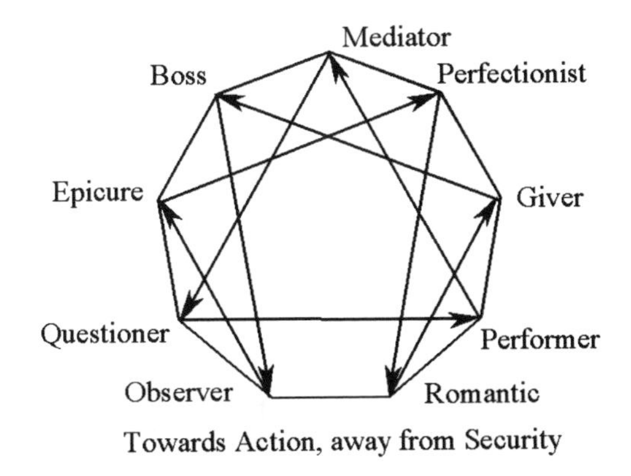

FIGURE 3 Character interactions.

empirical or, more likely, mystical. For example, the seven deadly sins, with two additions, were assigned to the personality types [1, p. 24]:

1	anger	2	pride	3	deceit
4	envy	5	greed	6	fear
7	gluttony	8	lust	9	sloth

Enneagramological writing is sometimes hard to distinguish from numerological writing. The following, about 2s, could have come from almost any numerology book

> Twos will merge identities with an authority and adapt into whatever the leader finds desirable. Although Twos have the capacities necessary to be a leader on their own, they generally prefer to be a power behind the throne, prime minister rather than king. From this strategic position, Twos will identify their own security with the authority's rise to power. By protecting the authority, Twos ensure their own future while at the same time earning love. It is unusual to find a Giver in an unpopular public position unless there is also an alignment to a power source. [1, p. 120]

There are more depths of enneagramism into which we will not go deeply. For example, the mind-body-spirit triad fits nicely into a nine-category system, as in Figure 4. If you are a 2, spirit is dominant. Your auxiliary triad, body, is the one closest to 2 and your "least integrated" triad, mind, is the one remaining.

Elaboration can be continued *ad lib.* For example [5], the numbers' communication styles are

1	moralizes	2	advises	3	propagandizes
4	laments	5	explicates	6	cautions
7	entertains	8	debunks	9	equalizes

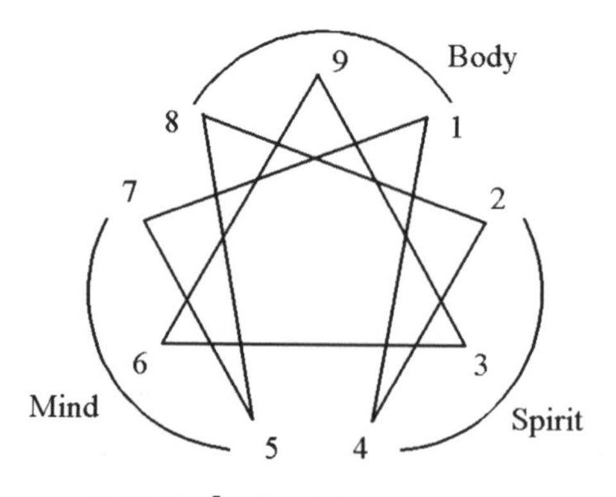

FIGURE 4 Mind, body, and spirit.

Their self-identities are

1	I am Hardworking	2	I am Helpful	3	I am Successful
4	I am Unique	5	I am Knowing	6	I am Loyal
7	I am Nice	8	I am Powerful	9	I am Okay

Enneagramism can probably do no harm, other than exacerbating peoples' tendency towards self-absorption. On the other hand, it is hard to see how it can do any good.

What enneagrams illustrate is, once again, the power of numbers. The numbers *do* something. They *add* something. A person at an enneagram conference was quoted in [2] as saying "I have to be more tolerant of eights; they just get on my nerves." It would be just as easy to say "bosses" as "eights," but the numericalization gives... something extra. The feeling of being on the inside, of being part of a special group, of having "knowledge" that outsiders lack? I think so: the *Newsweek* article reported that "Others snidely referred to acting spacey as 'Nining out.' "

We will have to wait to see what happens to enneagrams. Since they do not really provide anything new I predict that they will wither away. Enneagramists are taxonomists and after a time taxonomy palls. Phrenology, which divided people into classes according to the shapes of their skulls, was similar. The parallel is not exact, since phrenology pretended to have a physical basis, but it too told people about their character. While people like that, and easily can be convinced that they have almost any property (especially if it is a good one), it does not *lead* anywhere. I once heard a phrenologist lecture, but that was fifty years ago and the lecturer must have been one of the last extant. Phrenologists have disappeared. Enneagramists will too.

References

1. Palmer, Helen, *The Enneagram*, Center for the Investigation and Training of Intuition, 1988, HarperCollins paperback edition, 1991.
2. *Newsweek*, September 12, 1994, 64.
3. O'Leary, Pat and Maria Beesing, *The Enneagram: A Journey of Self Discovery*, Dimension Books, Denville, N. J., 1984.
4. ———, *Workbook for Enneagram: Basics*, Cleveland, Ohio, 1991.
5. Walker, Kenneth, *Gurdjieff, a Study of His Teaching*, Unwin, London, 1979.

CHAPTER 29

All that Glistens

Open a popular encyclopedia [6] to "golden section" and here is what you find:

> The concept of the golden section is of historical importance in aesthetics, art, and architecture. It has often been thought that a form, including the human form, is most pleasing when its parts divide it in golden sections. A related concept is the golden rectangle, which is a rectangle that has adjacent sides with lengths in the golden ratio. The ancient Greeks felt that the golden rectangle had proportions that were the most aesthetically pleasing of all rectangles; the shape appears in many works from antiquity to the present. It is especially prevalent in RENAISSANCE ART AND ARCHITECTURE.

Readers will tend to miss the cautious "it has often been thought" and conclude that the information in the paragraph is settled and authoritative. It's in the encyclopedia, isn't it?

Yes, it is in the encyclopedia, and it is wrong. A large amount of misinformation has become attached to φ, so tightly that it probably cannot ever be dislodged. (The golden ratio is

$$\varphi = (1 + \sqrt{5})/2 = 1.618\ldots$$

and Figure 1 is a picture of a golden rectangle.) It would be nice, and tidy, if the golden section unified aesthetics, art, and architecture, but life is more often messy than tidy. George Markowsky of the University of Maine, in a splendid article from which a good deal of the information in this chapter comes, has tried to explode some of the myths about φ [9]. He succeeded, but whether the truth will ever catch up with the errors is doubtful. The errors are much more satisfying and delightful than the dull and prosaic truth. Professor Markowsky says,

> The golden ratio... has captured the popular imagination and is discussed in many books and articles. Generally, its mathematical properties are correctly stated, but much of what is presented about it in art, architecture, literature,

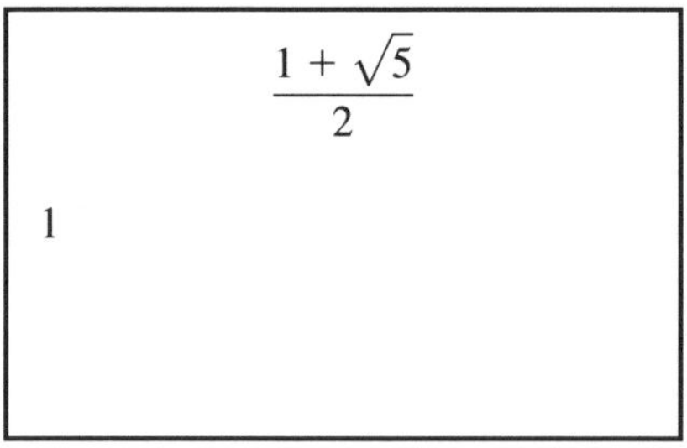

$$\frac{1+\sqrt{5}}{2}$$

1

FIGURE 1 A golden rectangle.

> and esthetics is false or seriously misleading. Unfortunately, these statements about the golden ratio have achieved the status of common knowledge and are widely repeated. Even current high-school geometry texts . . . make many incorrect statements about the golden ratio.
>
> It would take a large book to document all the misinformation about the golden ratio, much of which is simply the repetition of the same errors by different authors. [9, p. 2]

Let us first consider the myth that the golden rectangle is the prettiest of all rectangles. It is undeniable that some rectangles are prettier than others. In Figure 2, I am sure that everyone would choose the fourth of the four rectangles as the best-looking. The first is too . . . *flat*; the third is clearly going to fall over at any moment and hence makes us nervous; the second is square and squares are dull; the fourth, the golden rectangle, wins the beauty contest.

But the fourth would still win if it were not quite golden. One with a ratio of 3 to 2 would look better than the others, as would one with sides as 1.75 is to 1. Professor Markowsky says that the idea that the golden rectangle was better than any other got started in the 1860s, based on some experiments by Gustav Fechner:

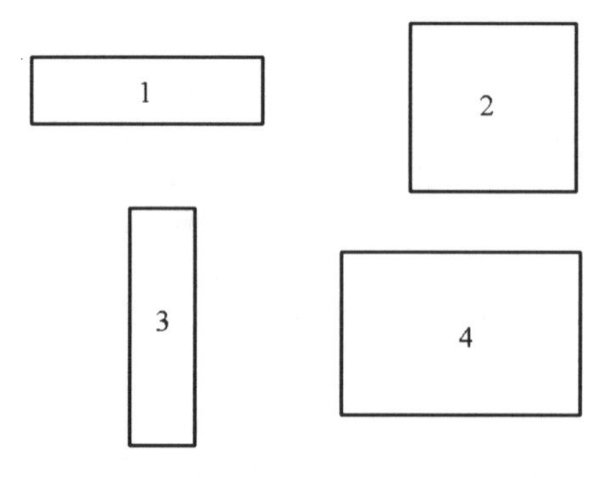

FIGURE 2 Find the prettiest rectangle.

> Fechner's procedure consisted in placing 10 rectangles before a subject and asking him to select the most pleasing rectangle. The rectangles varied in their height/length ratios from 1.00 (square) to .40.... The modal rectangle had a height/length ratio of .62, i.e., the golden section, with 76% of all choices centering on three rectangles having the ratios .57, .62, and .67. While all other rectangles received less than 10% of the choices each, Fechner's results still indicated that many other rectangles besides the golden-section rectangle were considered the most pleasing by a fair number of subjects. [9, p. 13]

A later study quoted by Professor Markowsky [10] concluded that

> Research on the golden section proportion as an empirically demonstratable preference has most often been applied to the rectangle where the results, on the whole, are negative.

Professor Markowsky conducted his own experiment. He constructed a six-by-eight grid of forty-eight rectangles, each with the same height and with lengths varying from .4 to 2.5 of the height. He included two golden rectangles, one with ratio $1/\varphi$ and one with ratio φ. He says,

> My informal experiments asking people attending my lectures to select the "most pleasing rectangle" suggest that people cannot find the golden rectangle in [the grid]. Furthermore, they generally select slightly different rectangles as the most pleasing rectangles....
>
> In the experiments I have conducted so far, the most commonly selected rectangle is the one with a ratio of 1.83....
>
> The various claims made about the esthetic importance of the golden ratio seem to be without foundation. [9, p. 14]

About the human form, there is the myth that the navel divides it, vertically, in the golden ratio. In his investigation of this claim, Professor Markowsky, understandably leery of asking strangers "May I measure your navel?" stayed within the home:

> While it might be entertaining to compute the ratio of many people's heights to the elevation of their navels, I did not spend much time on this effort. I did compute the ratios for the four members of my immediate family: 1.59, 1.63, 1.65 and 1.66. Their average is 1.63, which falls within our test interval for the golden ratio, although even in the small sample there is a significant amount of variation. [9, p. 15]

Professor Markowsky did not know of a more extensive study, since its results were reported at about the same time his paper was published. In [1], the authors report measuring the navel heights of 319 students at Middlebury College, finding that navels were consistently higher than the golden mean would have them (thus agreeing with the Markowsky family's navels) by about three-quarters of an inch. The results were highly statistically significant. Male navels tended to be slightly higher than female ones. The authors could not restrain themselves from writing "Admittedly, measurement error in this study was a ticklish problem" but, in view of the importance of their data, we will forgive them.

The encyclopedia excerpt quoted at the start of this chapter could give the impression that the ancient Greeks, impressed by the ubiquity and importance of φ, called it "the golden ratio." This is not so. This name is much more recent than that, as Professor Markowsky, quoting from [4, p. 14], points out:

> It may surprise some people to find that the name "golden section"... seems to appear for the first time in print in 1835 in the book *Die reine Elementar-Mathematik* by Martin Ohm, the younger brother of the physicist Georg Simon Ohm.... The first use in English appears to have been in the ninth edition of the *Encyclopaedia Britannica* (1875), in an article on Aesthetics by James Sully... The first English use in a purely mathematical context appears to be in G. Chrystal's *Introduction to Algebra* (1898). [9, p. 4]

At the end of the fifteenth century, Pacioli called φ "the divine proportion" in his *De Divina Proportione*. Divinity sounds more impressive than mere earthly gold, but the name has not survived, though Huntley's *The Divine Proportion* [8] preserves that name for φ, as well as many of the errors about it. Before Pacioli's time the only way to refer to the golden mean was as Euclid did, calling it "division in extreme and mean ratio."

Even if the ancient Greeks did not call the golden mean by any special name, many authors say that it is built into the Parthenon. Professor Markowsky says,

> To support this claim, authors often include a figure... where the large rectangle enclosing the end view of the Parthenon-like temple is a golden rectangle. None of these authors is bothered by the fact that parts of the Parthenon are outside the golden rectangle. [9, p. 8]

He goes on:

> Marvin Trachtenberg and Isabelle Hyman give the dimensions of the Parthenon as: height = 45 feet 1 inch; width = 101 feet, 3.75 inches; length = 228 feet 1/8 inch. They do not specify the points between measurements. These numbers give the ratios width/height $\approx 2.25 = 9/4$ and length/width ≈ 2.25 which are well outside the acceptance range. The reader might be struck by the fact that the ratio 2.25 appears as the ratio of width/height and length/width. Stuart Rossiter gives the height of the apex above the stylobate as 59 feet. This gives a ratio of $101/59 \approx 1.71$ which also falls outside the acceptance range. [9, p. 9]

It seems likely that the almost-golden rectangle in the Parthenon is coincidental.

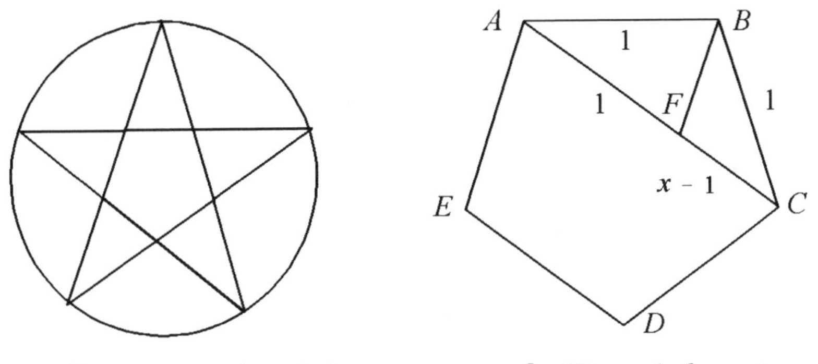

FIGURE 3 Pentagram in a circle.

FIGURE 4 Diagonal of a pentagon.

One reason for searching out φs in ancient Greece may be its mystical association with the Pythagoreans. Tradition has it that the emblem of the Pythagorean society was a pentagonal star inscribed in a circle, as in Figure 3. The figure contains quite a few φs. For one thing, the diagonal of a regular pentagon is φ times its side. In Figure 4, the pentagon has side 1. Draw the diagonal AC, with length x and make $|AF| = 1$. Then some angle-chasing shows that triangles ABC and BFC are similar:

$$\angle DEC = \angle CDE = 108^\circ$$

$$\angle EAC = \angle DCA = 72^\circ$$

$$\angle FAB = \angle FCB = 36^\circ$$

$$\angle AFB = \angle ABF = 72^\circ$$

$$\angle FBC = 36^\circ$$

$$\angle CFB = 108^\circ$$

so triangles ABC and BFC are both 36°-108°-36° triangles. Thus

$$\frac{BC}{FC} = \frac{AC}{BC} \quad \text{or} \quad \frac{1}{x-1} = \frac{x}{1}.$$

That is,

$$x(x-1) = 1 \quad \text{or} \quad x^2 - x - 1 = 0.$$

From this the quadratic formula gives

$$x = \frac{1 \pm \sqrt{1 - 4(-1)}}{2} = \frac{1 \pm \sqrt{5}}{2}.$$

Since $x > 1$, we have $x = \varphi$.

In the five-sided star in Figure 3, the ratio of the length of the extension of the sides to the points of the star to the length of the side of the inner pentagon is also φ. Yet further, if you make a new pentagon by joining the points of the star, the ratio of its side to the side of the small pentagon is φ.

The reason φ is everywhere is that the equations $x^2 + x - 1 = 0$ and $x^2 - x - 1 = 0$, having small coefficients, naturally arise in many different places. This is another application of the Law of Small Numbers. For example, consider the isosceles triangle circumscribing a semicircle in Figure 5. If A is far above D, triangle ABC will have a large perimeter, as it will if A is near the semicircle. Somewhere in between there will be a triangle with a minimum perimeter. It turns out to be the one with side $\varphi^{3/2}$, altitude φ, and perimeter $2\varphi^{5/2}$. The reason for this, as shown in [2], is that if the radius of the semicircle is 1 and the base angle of the triangle is θ, then its perimeter is

$$2(\tan\theta + \cot\theta + \csc\theta).$$

When we set the derivative of the perimeter equal to zero, out comes

$$\cos^2\theta + \cos\theta - 1 = 0,$$

from which φ emerges. If we put E at the point of tangency of the triangle to the circle, than triangle ADE is the golden right triangle, the one with sides 1, $\varphi^{1/2}$, and φ. The golden right triangle has not received the same notoriety as the golden rectangle, and I know of no claims that it is the most beautiful right triangle.

For another example of the commonness of φ, consider the fairly natural question of asking for which functions f, if any, it is true that

$$f(f'(x)) = x.$$

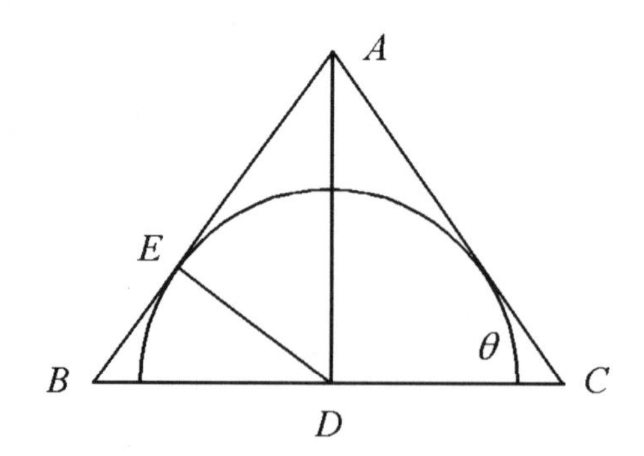

FIGURE 5 A φ minimum.

The answer is not known in general but, as is shown in [5], if $f(x) = x^n$, then $n = \varphi$. The author wrote,

> Is it not strange... that the value of n [is] the golden ratio $\frac{1}{2}(1 + \sqrt{5})$? This is the number to which the ratio of successive terms of the Fibonacci sequence tends. Perhaps there is a fundamental reason as to why the golden ratio should appear there.

The shallow, and sufficient, answer to that question is that the value appears because it is the number y that makes $y(y - 1) = 1$. Nothing deep is occurring, and that the question was even asked is evidence of the mystic appeal of φ. For the same reason, φ also appears when you ask for a power of x that makes $f'(f(x)) = x$.

It is also possible to find φ outside mathematics. Professor Monte Zerger, a keen observer, demonstrates in [11], using the best numerological principles, that Illinois is the golden state. The telephone area code 618 ($1/\varphi = .618\ldots$) and the postal zip codes that start with 618 both have been assigned to Illinois. Area code $309 = 618/2$ is also in Illinois. If you cut the contiguous forty-eight states west to east in the golden ratio, the cutting meridian, $89^\circ\ 1'$, passes through Illinois. Do the same south to north and the golden parallel of latitude, $39^\circ\ 54'$, also passes through Illinois. The two lines intersect near Decatur.

Since φ is also approximated by ratios of successive Fibonacci numbers, it is also highly significant that

> Illinois was admitted to the Union on the *3*rd of the month (December, 1818).
>
> Illinois is the *5*th largest state in order of population (1980 census).
>
> Illinois has *8* letters.
>
> Illinois is the *13*th state in alphabetical order.
>
> Illinois was the *21*st state to join the Union. Its post office abbreviation, IL, is composed of the ninth and twelfth letters of the alphabet, and $9 + 12 = 21$. φ is the twenty-first letter of the Greek alphabet.
>
> Interstate highway *55* begins in Illinois and stays close to longitude *89*.
>
> Longitude *89* cuts the country in the golden mean, and the barge canal connecting Lake Michigan with the Mississippi River gives (with the Great Lakes) the only water cut in the country.

Professor Zerger evidently had no luck in finding a 34 connected with Illinois. U.S. Highway 34 runs from Chicago to the Mississippi River, and it looks as if at some point it might divide the state north to south in something like the golden section. More research is needed, not only for 34, but for 144, 233, 377, . . .

All of those Fibonacci numbers occur because of the inexorable workings of the Law of Small Numbers and, as Professor Zerger knew, have no other

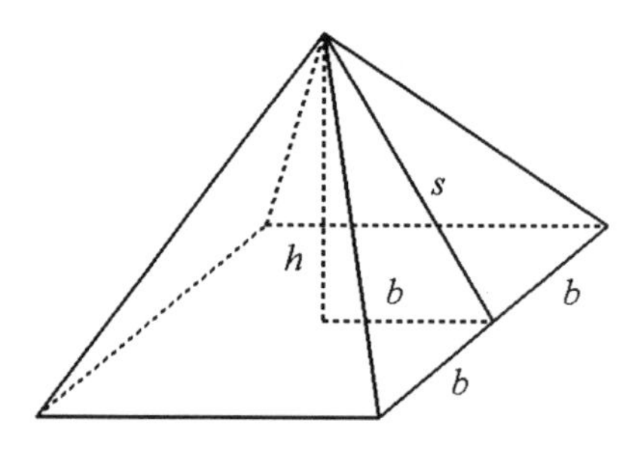

FIGURE 6 A square pyramid.

significance. Numerologists, on the other hand, might read into them all kinds of mystical significance. They might take them as proof that the builders of the great pyramid of Egypt emigrated to Illinois after they finished.

The pyramid also has φ-associations. It is a fact that the ratio of its slant height, s in Figure 6, to half the base, b, is almost φ, but that was explained in chapter 24 to be the result of the slope that the pyramid builders chose.

The story has gotten around that Herodotus was told that the pyramid was constructed so as to embody φ. Professor Markowsky located the following story in several sources, including a well-regarded history of mathematics:

> Herodotus related in one passage that the Egyptian priests told him that the dimensions of the Great Pyramid were so chosen that the area of a square whose side was the height of the great pyramid equaled the area of a face triangle. [9, p. 7]

Were this the case, we would have $h^2 = bs$, which, together with $b^2 + h^2 = s^2$ (from the Pythagorean theorem) gives

$$b^2 + bs = s^2 \quad \text{or} \quad 1 + \left(\frac{s}{b}\right) = \left(\frac{s}{b}\right)^2,$$

from which the φ would come.

There is nothing like this in the works of Herodotus. What he had to say about it is, in the translation that Professor Markowsky used,

> The Pyramid itself was twenty years in building. It is a square, eight hundred feet each way, and the height the same, built entirely of polished stone fitted together with the utmost care. The stones of which it is composed are none of them less than thirty feet in length. [9, p. 7]

Roger Herz-Fischler [7] has traced the story about making the square equal in area to the triangle to the original pyramidologist, John Taylor, in his 1859 book *The Great Pyramid*. Taylor either made the story up or, more likely, arrived at it by an eccentric interpretation of Herodotus's original Greek. As Professor Markowsky says,

> The distorted version of Herodotus's story makes little sense. Even the authors who quote it do not give a reason why the Egyptians would want to build a pyramid so that its height was the side of a square whose area is exactly the area of one of its faces. The idea sounds like something dreamt up to justify a coincidence rather than a realistic description of how the dimensions of the Great Pyramid were chosen. It does not appear that the Egyptians even knew of the existence of φ much less incorporated it in their buildings. [9, p. 7]

A question remains: Why did Herodotus get the dimensions of the pyramid wrong? Its base is about 755 feet square, but its height, about 480 feet, is far short of 800 feet. I have an answer. I have had the privilege of gazing on the great pyramid. If you had asked me, while I was gazing, how the length of the base and the height compared, I would have said that they looked about the same. And so they did, and do. Vertical distances impress themselves on the mind with more force than horizontal ones. One hundred yards on the flat is nothing, merely the length of a football field. But a building three hundred feet high is so tall that it's a skyscraper. Similarly, people think that slopes are steeper than they really are. A hill rising at an angle of twenty degrees impresses itself on the eye as tremendously steep. If you asked a pyramid-gazer at what angle its sides were rising, I expect that you would get an estimate in the vicinity of seventy-five degrees rather than one close to the correct fifty-odd degrees. Herodotus was no mathematician or engineer and, as with many historians, it is possible that he did not care much about numbers and wrote what he did by following the impression that his eyes gave him.

There is no φ in the pyramid. The Egyptians were too primitive even to know about it. It was refreshing to read the following in a popular novel:

> At the university they had a lot of funny ethnological clichés. One of them was about how much European mathematics was indebted to ancient folk culture; just look at the pyramids, whose geometry commands respect and admiration.
>
> This, of course, is idiocy disguised as a pat on the back. Technological culture is superior in the very reality it defines. The seven or eight rules of thumb of the Egyptian surveyors is abacus mathematics compared to integral calculus. [7, p. 188]

Finally, the stories that φ appears in the United Nations building (or in architecture generally), in the paintings of da Vinci, Mondrian, and Seurat, and in the structure of Virgil's *Aeneid* are just that—stories, based on someone's observation of a rectangle that looked golden, or of some Fibonacci numbers. It would be pleasant if we had discovered in φ the grand unifying principle of nature, since it would be nice to live in a tidy and comprehensible world, but we have not. Alas, all is not golden.

References

1. Calise, Lori K., John B. Cunningham, Tamara M. Caruso, and Paul M. Sommers, The golden midd, *Journal of Recreational Mathematics* **24** (1992) #1, 26–29.
2. DeTemple, Duane W., The triangle of smallest perimeter which circumscribes a semicircle, *Fibonacci Quarterly* **30** (1992) #3, 274.
3. Fischler, Roger, What did Herodotus really say, *Environment and Planning B* **6** (1976), 89–93.
4. Fowler, D. H., A generalization of the golden section, *Fibonacci Quarterly* **20** (1982), 146–158.
5. Gattei, Pierino, The "inverse" differential equation, *Mathematical Spectrum* **23** (1990/91) #4, 127–131.
6. *The Software Toolworks Multimedia Encyclopaedia*, version 1.5, Grolier Electronic Publishing, 1992.
7. Høeg, Peter, *Smilla's Sense of Snow*, translated from the Danish by Tiina Nunnally, Farrar Strauss and Giroux, New York, 1993.
8. Huntley, H. E., *The Divine Proportion*, Dover, New York, 1970.
9. Markowsky, George, Misconceptions about the golden ratio, *The College Mathematics Journal* **23** (1992) #1, 2–19
10. Schiffman, H. R. and D. J. Bobko, Preference in linear partitioning: the golden section reexamined, *Perception & Psychophysics*, **24** (1978), 102–103.
11. Zerger, Monte, The golden state—Illinois!, *Journal of Recreational Mathematics* **24** (1992) #1, 24–26.

CHAPTER 30

Numbers, Numbers Everywhere

Number mystics are people who have numbers in their heads and find that what they do there is valuable to them. Numerologists are people who also have numbers on the brain but are unable to keep them there: they must get out and *do* something.

E. O. provides an example of an amateur numerologist. He was very active in the 1970s in distributing his work far and wide. Though active, he was devoted to the passive voice:

> Original papers developing a theory of number that sees all fundamental numerical relations in the primary integers in the form of flow have been prepared and sent where interest in such a theory might be expected. It is hoped that the papers will be noted with sympathy and understanding. It would be appreciated if the papers might also be posted for the attention of others for their possible interest. If the papers are of no interest to anyone it may be so indicated.

He wrote

> It is proposed that number and nature are an indissoluble harmonic unity of integer-particles. Thus:

1	2	6
unity	motion	flow
3	5	7
mass	energy	power
4	8	9
form	force	harmony

This is pure mysticism. Note the similarities, or lack of them, between Mr. O.'s characteristics of digits and those of the modern numerologists. There are coincidences at 1 (not surprising) and 5, but not elsewhere. There are also differences from the enneagramatologists' characteristics. There is no unanimity among mystics. The conclusion is that integers do not have inner essences and vibrations that those sensitive to such things can sense.

Mysticism led Mr. O. to a value of π different from 3.1415926535.... Here is his new value and what he said about it. I quote at length, both to show the reasoning of number mystics and because I doubt that anyone else has bothered to preserve Mr. O.'s works. He wrote prolifically, producing his *Bulletins* every week or so. A typical *Bulletin* would be a one-page mimeographed sheet, though some ran to two pages. Since the title of Mr. O.'s π-sheet is "A Manifesto to Science," he evidently thought that it was important.

> A state of potential revolution in science is declared to exist in the concept of number presented herein that sees the primary numbers as a form of flow that is basic to all number and that is directly definable in terms of the primary integers. Such flow is seen and represented thus:

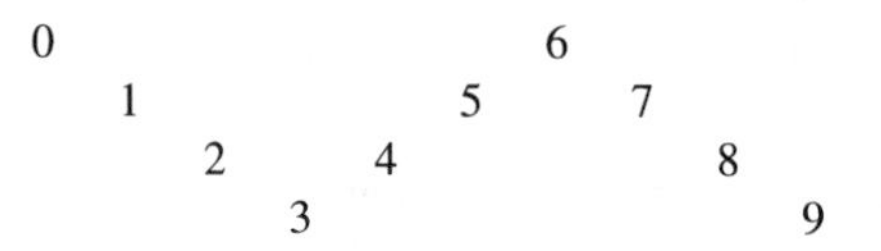

> The concept defines number in terms of the primary integers so represented as relations of numbers constituting fundamental values common to all numbers. Primary values so derived correspond to a high degree to the mathematical values of *e* and pi. The residual differences are evidence of contrary approaches which may be defined as direct and indirect, as intuitive and inventive. Thus the premises are fundamentally different.
>
> The new approach to number is traced in simple logic:
>
> We see 1 as the additive number and as the prototype of an additive order beginning with 1 5 7 as a unit of such order.
>
> We see 1 5 7 as dividing from 0 in an additive relation of 1 5 7 to define: an additive sum, 1; a multiplicative sum, 5; and a ratiotive sum, 7. We see 1 5 7 as a form of flow, 1 5 7.... We define 1 5 7... as an additive relation of the additive order.
>
> We see 1 5 7, further, as dividing from 0 in a multiplicative relation of 5 and 7 to define: an additive sum, $5 + 6 + 7 = 18$; a multiplicative sum, $18 + (18/2) = 27$; and a ratiotive sum, 2718. We see 2718 as a form of flow, 27182718.... We indicate value relative to 1 by a decimal point, 2.7182718.... We define the derived value as a multiplicative relation of the additive order. We compare the derived value with the mathematical value *e*, 2.7182818+.
>
> We see 1 5 7, finally, as dividing from 0 in a ratiotive relation of 1 5 7 to define: an additive sum, 157; a multiplicative sum, 314; and a ratiotive sum, 314157. We see 314157 as a form of flow, 314157314157.... We indicate value relative to 1 by a decimal point, 3.14157314157.... We define the derived value as a ratiotive relation of the additive order. We compare the derived value with the mathematical value pi, 3.14159265+.

> The new values for e and pi so derived are unique and have a validity of their own as derived from the nature of number itself at its source in the primary integers, than which there is no higher source of numerical knowledge. In fact, one may say that these values represent number in the only way in which number can be truly represented and in the only way in which the nature of number can be truly defined.
>
> The values so derived are there to be seen by those who will and they cannot be dismissed lightly. They stand in stark contrast to standard values and present a challenge of choice. They cannot be compromised and they cannot be correlated. The one set or the other must prevail. A different approach to number and a different philosophy of knowledge are represented by each set.

Numerology at its purest! The contemplation of numbers, and *only* the contemplation of numbers, without reference to anything outside, leads to the correct values of π and e.

But π and e are not conjurations of the mind: they are out there, in physical reality. Circles exist, and circles have the last say about what is the ratio of their circumferences to their diameters, no matter what Mr. O. or anyone else may think, and the value of e is fixed forever when we say that it is the number such that e^x has itself for its derivative.

Mr. O. knew that his values were not those that everyone else used. A choice must be made, the choice between mysticism and physics. Guess which won.

> It will not be easy to choose even if one faces the reality of the choices squarely. The prudent decision will be to ignore the new values, if one can, and to content oneself with the standard and established values which have worked "well enough" and have the weight of long tradition behind them. But can the witness of the primary integers be ignored, especially by those who proclaim "Truth above everything" as their way of life?
>
> A new number theory and a new mathematics will evolve naturally as the new values and their implications are understood. Particularly, the spirit of the new values as approached by direct intuition in the primary integers will lead to a new attitude toward number and nature. No more will one speak with arrogant presumption of "forcing nature to yield her secrets," which is but the language of idiots and maniacs, for it will be known that the primary integers do not yield the secrets of their nature to such violence. Instead, one may expect a new reverence for nature, a greater respect for intuition and faith, and a deeper awareness of intelligence as sensitivity and responsiveness to an intelligence that forever flows in the form of nature.

"As approached by direct intuition": there we have the mystic speaking, there we have the mystical approach to truth. And there we have the mystic going too far, when he asserts that intuition is superior to reason in determining the values of e and π. Intuition is superior to reason, sometimes (or is at least a more rapid means of getting at some forms of truth, emotional ones especially), but not for e and π, not for mathematics, not for science, not for getting at the physical truths of the universe. Mysticism does not provide information about physics. That, of course, is only an assertion. A mystic could answer, and many have, with a counterassertion, "You are wrong. Intuition leads to higher truths." There is no reasoning with mystics. You must affect their intuitions. I do not know how to do that.

When I wrote to Mr. O., saying that some of his reasoning was hard to follow and mentioning that I didn't think that mysticism was a good way of getting at the value of π, his reply included

> As to mysticism, I abhor it if it implies delight in mystery. My only delight is in the knowable. To me, mathematics is the ultimate mysticism, floating in thin air with absolutely no basis in a fundamental theory of number.... I am sure that Gauss' contemporaries found his "Disquisitiones" also quite incomprehensible.

After that, his production of *Bulletins* increased sharply for a time. He needed, I think, to restore his equilibrium. It had been disturbed.

I don't think that Mr. O.'s efforts convinced anyone, partly because he so often lapsed into incoherence. Here is the entire content of his *Bulletin* of July 22, 1974. Note the direct restatement of the Pythagorean principle that all is number in his first sentence. For all I know, Mr. O. arrived at the Pythagorean view independent of Pythagoras, Neopythagoreans, or modern Pythagoreans.

> THE NUMERICAL UNITY OF THE ATOM
>
> It is proposed that number *is* nature in its purest form, that the contemplation of number is the purest observation of nature, and that the harmonic values of number are ± 0 values of nature. Thus, the following numerical ratio is proposed as a pure prototype of nature, more direct and precise than any apprehension of the senses or of any photograph:
>
> $$(1\,/\,568\ 568\ 568\ 5)\,/\,1\ 756\ 765\ 567 = 100\ 115\ 9660$$
>
> The above ratio proposes that number evolves from a placid continuity of sameness, to a dynamic discontinuity of difference by a permutational evaluation of 7, to harmonic constancy by progressive cycles of unity in a conservation of unity. The process is proposed to be defined in particular in the composition of 1 756 765 567 as a progression toward the constancy of 567.
>
> Transposed into the physicality of nature, it is proposed that nature evolves in the atom in the same fundamental progression from a charge to mass ratio for the electron (1 / 568 568 568 5) to the magnetic moment of the electron (100 115 9660) by the evolution of the atom as a unit of nature (1) by permutation as proton (756), nucleon (765), and electron (567).

The dimensionality of number is proposed as follows, again in direct and precise prototype of nature:

$$5^3 \times \sqrt{6} \qquad 1913^3 \times \sqrt{265} \qquad 139\ 6162\ 654$$

$$\times \frac{5^3 \times \sqrt{6}}{7^2} \qquad \times \frac{1913^3 \times \sqrt{265}}{305^2} \ /\ 139\ 45\ 45\ 45\ 45$$

$$= 1913\ 265\ 305 \qquad = 139\ 6162\ 654 \qquad = 100\ 115\ 9660$$

It is not easy to see what to make of that. The calculations in the last line are correct (except my calculator gives 649 instead of 654 for the middle number) if you remember that Mr. O. did not bother with decimal points. The point, I think, was to make a mystic connection between the two 100 115 9660s. But it must have been meant to be more than mystic, it was meant to be physical, for there is all that talk of mass ratio and magnetic moments.

Here is one final example, to show the dangers and rewards of contemplating number. This is the *Bulletin* of August 23, 1973:

THE UNITY OF 1

It is proposed that 1 is a unity of 1 in itself but that 1 evolves as a unity of 1 in sums of 1 by ratios of sums of 1. Thus, a simple unity of 1 evolves in ratios of sums of 1 as 1 111 111 111 and a complex unity of 1 evolves in ratios of sums of 1 as $1\ 111\ 111\ 111^1$, $1\ 111\ 111\ 111^2$, and $1\ 111\ 111\ 111^3$.

It is proposed that the sum of $1\ 111\ 111\ 111^2$ evolves further as a principle of unity in a simple compact ratio defining a simple mean of 1 in 123 in the simple ratio $(1 + 3)/2 = 2$. A constant mean value of 1 in 123 456 789 0 is proposed in which the unity of 1 is conserved by a more complex ratio:

$$\left(1^3 \times \sqrt{2} \times \frac{1^3 \times \sqrt{2}}{3^2}\right)^2 = 2\ 222\ 222\ 222^2 = 49\ 38\ 27\ 16\ 05.$$

The above ratio is proposed as the fundamental form of the complexity of nature in its utmost simplicity. As such, it is proposed as the prototype of all of nature in its numerical composition as a ratio of sums of 1 in mean constancy and in dynamic equilibrium as a system of three-body systems. The sum of 49 38 27 16 05 as a formulation of the primary integers in a reciprocal spiral of 43210 and 98765 is proposed as fundamental in nature, whether in the structure of a spiral galaxy or a sunflower.

The sum of 49 38 27 16 05 as the primary structure of nature is proposed particularly as inherent in the structural fundamental constants and in the structural wavelengths of light by the very simple ratio of 49 38 27 16 05 in which they evolve.

I had never known, or perhaps it had never sunk in properly, that the digits of $4/81$ interlaced 43210 and 9876 so neatly:

$$4/81 = .04938271604938271604 9\ldots$$

That discovery is a reward of number-contemplation. The remainder of the *Bulletin* exhibits the danger: numbers can get into the head and swirl around until all is confusion. One wonders how many of the ancient Pythagoreans were people like Mr. O. I'll bet that there were quite a few.

Mr. O. did have some responses to his work, as the quotations in the following excerpt show. They also show the mystical reaction to mathematicians' criticisms.

THE SIMPLICITY OF NUMBER

"This is not mathematics, this is elementary arithmetic."

"Very few mathematicians think this way."

The above quotations serve to define the human problem of number, both for editors and for writers. Nevertheless, number by its simplicity is proposed to reveal reality that number in its mathematical sophistication cannot see, perhaps even when shown. Thus, the "absurd" simplicity of the numerical cycle presented below proposes to define the inherent and potential proportions of the universe in a positive progression of cycle ratios of reciprocal sums and ratios of 31 and 32:

$$12^3 \times \sqrt{31} \times \frac{12^3 \times \sqrt{31}}{27^3} = 126\,976\,0000$$

where 126 976 0000 =

31×4096	32×3968	203 1616 / 16
62×2048	64×1984	406 3232 / 32
124×1024	128×992	609 4848 / 48
248×512	256×496	812 6464 / 64
496×256	512×248	1015 8080 / 80
992×128	1024×124	1218 9696 / 96

That was a new idea to me, that mathematics has corrupted the primeval simplicity of number by its sophistication. Poor, ravished number!

The question is, should mystics and mathematicians try to communicate with each other? *Can* they communicate? I think that their worlds are too far apart.

CHAPTER 31
Biorhythms

There are cycles in nature, not just the obvious ones of the seasons of the year, the phases of the moon, the ebb and flow of the tides, or the appearance of seventeen-year locusts, but others whose causes are not yet completely understood, such as the eleven-year sunspot cycle. There are also cycles in people. Besides the obvious ones such as waking and sleeping, some hormones vary in concentration, up and down, regularly and predictably.

This is no news. But cycles have numbers attached to them, and they are patterns. Some humans have the ability to see patterns where there are none (for example, see the next chapter) and others can become fascinated with numbers. When one person has both tendencies, bad things can happen. An example is given by the originator of biorhythms, Wilhelm Fliess (1858–1928), doctor and friend of Sigmund Freud, whose letters to Fliess make up a whole book [1]. Fliess took to the idea of cycles and managed to persuade other people susceptible to cycles and numbers that his baseless ideas had some worth. Such is the power of numbers.

Fliess found that several illnesses of the body could be alleviated by the application of cocaine to the nose. No doubt his patients *said* that they felt better. He then became interested in the nose, studied it closely, and found that it underwent cyclic changes, with a period of twenty-eight days. Twenty-eight days, the same length as the menstrual cycle! Clearly, here is a discovery.

Somehow he got the idea that there was also a twenty-three day cycle operating in people.

> The facts before us compel us to emphasize another factor. They teach us that, apart from the menstrual process of the twenty-eight day type, yet another group of periodic phenomena exists with a twenty-three day cycle, to which people of all ages and both sexes are subject. [1, p. 7]

But the facts before us were not spelled out, and the origin of the twenty-three is obscure. The only indication I found was the following:

> From the nasal findings that the menstrual intervals from July varied between twenty-three and thirty-three days. [1, p. 158]

He made his discovery in 1897 and his book about it, *Der Ablauf des Lebens* (The Course of Life), appeared in 1906, with a second edition in 1923. Martin Gardner gave an account of Fliess, Freud, and his cycles in one of his *Scientific American* columns, reprinted with additions in [2, 3]. He says that Fliess's book

> is a masterpiece of Teutonic crackpottery. Fliess's basic formula can be written $23x + 28y$ where x and y are positive or negative integers. On almost every page Fliess fits this formula in natural phenomena, ranging from the cell to the solar system. The moon, for example, goes around the earth in about 28 days; a complete sunspot cycle is almost 23 years.
>
> The book's appendix is filled with such tables as multiples of 365 (days in the year), multiples of 23, multiples of 28, multiples of 23^2, multiples of 28^2, multiples of 644 (which is 23×28). In boldface are certain important constants such as 12,167 [23×23^2], 24,334 [$2 \times 23 \times 23^2$], 36,501 [$3 \times 23 \times 23^2$], 21,952 [28×28^2], 43,904 [$2 \times 28 \times 28^2$], and so on. A table lists the numbers 1 to 28, each expressed as a difference between multiples of 28 and 23 [for example, $13 = (21 \times 28) - (25 \times 23)$]. Another table expresses numbers 1 through 51 [23 + 28] as sums and differences of 23 and 28 [for example, $1 = (\frac{1}{2} \times 28) + (2 \times 28) - (3 \times 23)$]. [3, pp. 154–155]

The tables demonstrate two things. First, the fascination of numbers and second, Fliess's lack of mathematical knowledge. The fascination, or perhaps respect, is implicit in the existence of the tables ("Look! Look at the numbers! Aren't they wonderful?"), and the lack of knowledge is shown by his not knowing that $n = 23x + 28y$ can easily be solved for x and y. For example, to find x and y so that $23x + 28y = 2000$, look at the equation modulo 23:

$$5y \equiv 22 \equiv 45 \pmod{23}, \qquad y \equiv 9 \pmod{23},$$

so $y = 9$ will work. Then $x = (2000 - 9 \cdot 28)/23 = 76$. Even better, there are infinitely many solutions, $(x, y) = (76 - 28t, 9 + 23t)$ for any integer t. Furthermore, the largest integer that cannot be written in the form $23x + 28y$ with both x and y positive is 593:

$$23 \cdot 10 + 28 \cdot 13 = 594$$

$$23 \cdot 21 + 28 \cdot 4 = 595$$

$$23 \cdot 4 + 28 \cdot 18 = 596$$

$$23 \cdot 15 + 28 \cdot 9 = 597$$

and so on for 598, 599, ..., but the best that can be done for 593 is

$$23 \cdot (-1) + 28 \cdot 22 = 593.$$

But I am letting myself get carried away with the fascination of numbers. I wonder what Fliess would have made of that 593. I'm sure he could have given it *some* significance.

The idea of the theory is that at the moment of birth a person's cycles are started in motion, oscillating from the starting point of zero up to a maximum, back down through zero to a minimum, then returning to zero to repeat, periodically, until death. Why the cycles do not start at the moment of conception—there seems to be no reason why fetuses should not also have ups and downs, good days and bad—is not explained. The reason, of course, is that the day of birth is recorded and the day of conception unknown.

Later Fliessians added a 33-day cycle to Fliess's 28- and 23-day cycles, so that each person has three waves. I suppose that 33 was chosen for symmetry's sake, so that 28 is in the middle. Figure 1 shows the cycles during the first 33 days of life.

The addition of the third cycle may have been prompted by difficulties that arose with the two original ones:

> But Fliess soon found himself obliged to explain the intervals with which he was confronted by combination of four figures and to use not only twenty-three and twenty-eight, but five (28 − 23) and fifty-one (28 + 23).
>
> [1, p. 40]

Since the 28-day cycle, occurring as it does in menstruation, is feminine (note that 28 is an even number, and hence female), the 28-day cycle is the emotional cycle. Since 23, being an odd number, is male, it is appropriate that the 23-day cycle measures physical ups and downs. So, we have two of the famous body-mind-spirit triad accounted for, whence the third cycle must complete it and so is the intellectual cycle. As is usual with number mystics, no reasons are given for all these assertions or for the choice of 23, 28, and 33. Not only is it the case that not all knowledge comes from reason, if you do not give reasons you avoid many tiresome arguments.

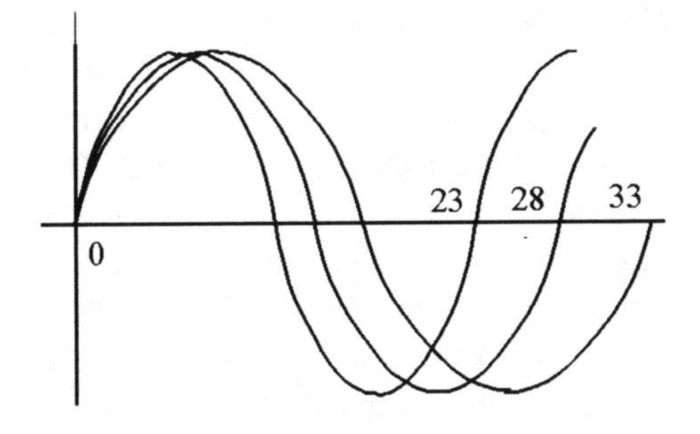

FIGURE 1 The three cycles.

Once you have your cycle chart, its interpretation is obvious: do not try to prove difficult theorems when at the bottom of the intellectual cycle, try no record high jumps while at the bottom of the physical cycle, and avoid confrontations in relationships while at the bottom of the emotional cycle. Contrariwise, when at a physical, emotional, or intellectual peak, successes may be achieved that on ordinary days would be impossible. Double peaks are at least doubly good, and triple tops—those are the days! The trouble is, they do not come very often. In fact, the 23- and 28-day cycles never peak simultaneously. (The 23-day cycle has a peak at time 5.25 and every 23 days thereafter, while the 28-day cycle peaks at time 7 and then again every 28 days. If both had a peak at the same time, we would have

$$5.75 + 23r = 7 + 28s$$

or, multiplying by 4,

$$23 + 92r = 28 + 112s.$$

But the left-hand side is odd and the right-hand side is even and so are never equal.) However, there are many days when both are near their peaks. Look at your chart and you will find them.

Because 23, 28, and 33 have no factors in common, all three curves will not reunite at zero until $23 \times 28 \times 33$ days have passed. That is 21,252 days, or a bit more than 58 years. Thereafter, you and everyone else will have the same ups and downs that you previously experienced, though after that lapse of time you probably will not be able to remember how you felt, thought, and acted the first time around.

Days especially to be watched out for are those where one of your curves crosses the x-axis. As with the rest of the theory, there is no more reason behind this assertion than the assertions of other forms of numerology. That is, it is mystical: days when the curve cuts the axis are days when one of the characteristics changes sign and where the slope of the curve is at a maximum. That is, it is a day of severe change, and days like that are to be treated with caution. If there are two crossovers, then the danger is at least doubled, and with three it is even worse, though whether tripled, quadrupled, or octupled (2^3) I do not think the biorhythmists have decided. In any event, if it has not yet come, watch out for that day in your 58th year when all three curves simultaneously plunge toward the axis, reaching zero on the way down for the first and last time in your life. Unless you survive to be 116, that is.

There is no reason to think that Fliess was a Pythagorean of any kind, but he echoed, or came close to echoing, the Master's doctrine of the music of the spheres, as well as the mystical All is One:

> ... the periods then continue in the child, and are repeated with the same rhythm from generation to generation. They can no more be created anew than can energy, and their rhythm survives as long as organized beings reproduce themselves sexually. These rhythms are not restricted to mankind, but extend into the animal world and probably throughout the organic world. The wonderful accuracy with which the period of twenty-three, or, as the case may be, twenty-eight whole days is observed permits one to suspect a deeper connection between astronomical relations and the creation of organisms.
> [1, p. 7]

One of Freud's patients, Hermann Swoboda, took up Fliessian biorhythms and kept the movement alive. For some reason, Germany in the 20s and 30s was fertile ground for pseudoscience and mysticism. For example, the idea that the earth is hollow flourished there and then as nowhere else and at no other time. There are not many hollow-earthers left now (though there are some, ready to produce details of the seven-foot high beings who live inside Mt. Shasta), but there are still plenty of biorhythmists. The reason may be that, unlike the hollow earth, biorhythms can be used for prediction and, as with astrology, there are people who will believe the predictions. Whatever they are. Martin Gardner tells how the magician James Randi played an unkind trick that demonstrated, yet again, that some people have the capacity to believe almost anything.

> After one of his shows, Randi tells me, a lady in New Jersey sent him her birth date and asked for a biorhythm chart covering the next two years of her life. After sending her an actual chart, but based on a *different* birth date, Randi received an effusive letter saying that the chart exactly matched all her critical up and down days. Randi wrote back, apologized for having made a mistake on her birth date, and enclosed a "correct" chart, actually as wrongly dated as the first one. He soon received a letter telling him that the new chart was even *more* accurate than the first one. [3, pp. 158–159]

Biorhythms had an upsurge in the late 70s. There were articles on the subject in popular magazines and it was possible to buy biorhythm books, kits, and calculators. One result of this efflorescence was that it caught the eye of scientists, some of whom decided to put biorhythms to the test. Biorhythms failed. In "Biorhythms and tests of oral skills" [4], no relation of cycles to oral skills were found. In "Biorhythms and their effect on patient expectation in hypnosis" [5], data on 34 suicides, 500 automobile accidents, and 220 oil-drilling accidents showed no relation to biorhythm cycles. In addition,

> It is concluded that patient belief and dependency on biorhythm cycles are as inhibiting to developing responsibility and self-control as other misinformation and superstitions.

In Romania, 632 accidents gave no support to Fliessian biorhythms, nor did the results of 98 students taking intelligence tests on two different days [6]. The psychiatric emergency room visits of 604 patients were randomly distributed among cycle days [7]. So were 215 labor accidents in Romania [8]. In Australia, 1239 accidents in a timber-milling organization were random [9], as were climbing accidents in the Grand Teton National Park [10]. Accidents in Poland [11], 199 collegiate football injuries [12], 881 motor vehicle accidents [13], 2241 deaths in *Webster's American Biographies* [14], the psychiatric hospitalization of 218 patients [15], landing performance by pilots [16], the performance of 610 European track and field athletes [17], the performance of participants in an archery league [18]: all were independent of Fliessian biorhythms.

I could go on for quite some time citing other studies (382 homicides in Philadelphia in 1982, 70 major league baseball players' performance in 1975, 400 mining accidents in Canada, and so on) but I will not. For mathematicians, one counterexample is enough to show a theorem is false and once one is found there is no need for anyone to look for another. To be sure, testing statistical hypotheses at the 5% level is different from trying to prove theorems, since we can never have 100% certainty. Rejecting the hypothesis at the 5% level does not mean that it is false, only that the probability of its truth is less than .05. But when the number of rejections climbs from ten towards twenty, that should be enough to put it to rest. After all, $(.05)^{10} = .00000000000009765625$, a probability that should be small enough for anyone to reject anything.

However, we must be fair and admit that the returns are not unanimous against biorhythms. Three studies found something in them. The first investigated 58 deaths from heart attacks from 1977–79 in Cuba [19]. It concluded that

> for Ss as a whole, the first symptom of MI [myocardial infarction] generally coincided with the critical phase of the physical cycle.

Another study of 40 MI cases in 1981 [20] showed that

> in the majority of cases, the acute infarction occurred during the negative phase of the biorhythmic cycle.

This appeared in the *Boletin de Psicologia Cuba*, the same journal that printed the first study. The final paper [21] looked at 29 women whose pregnancies ended with the birth of a dead fetus:

> A highly significant and frequent correlation was found between the delivery date of the dead fetuses and the negative phase of the physical cycle of the mothers' biorhythms.

The journal that *that* appeared in was the *Revista del Hospital Psiquiatrico de La Habana*. I don't need any statistical test to conclude that someone in Cuba thought very highly of Fliessian biorhythms. The probability is 1.

By the way, so as not to give a false impression of my erudition, I should admit that I have not seen the originals of any of these studies. However, the miracle of modern database technology allows abstracts of them to be found very quickly.

True believers like the doctor in Cuba can keep biorhythms alive, like viruses lying dormant in the body, until their next outbreak. Though they have subsided for now, and the last computer program that I saw advertised that would determine your biorhythm cycle was drastically marked down to $12.95, they have not gone away. The American Psychological Association database from which almost all of the preceding examples were taken contains in addition an abstract of a 1991 paper, "Belief in the paranormal: a New Zealand survey" [22], that asserted that over 30% of 1048 university students expressed belief in biorhythms. It remains to be seen if this percentage is on the ascending or descending part of its cycle curve. (If biorhythms hold for people, why shouldn't they hold for doctrines as well? As Fliess said, "These rhythms are not restricted to mankind.") If descending, it also remains to be seen if biorhythms will disappear when the cycle hits one of its bottoms. We can hope so, since idiocies sometimes die, but it also possible that biorhythms will always be with us. Such is the power of numbers.

It is easy to make your own biorhythm chart, even without a $12.95 computer program. All that is necessary is to calculate the number of days, D, since your date of birth. Once you have that, your position on your three cycle curves is determined by D modulo 23, 28, and 33, which can be calculated quite quickly. So can D: years of your life times 365, plus the number of February 29ths you have lived through, plus the number of days from your last birthday. I find that today I am in day 1 of the physical cycle, day 8 of the emotional, and day 5 of the intellectual. I am one day past my emotional peak, so it is not the case that every day in every way I am getting better and better, but nevertheless, everything is positive. That is nice to know. It almost guarantees that I will not die of a heart attack in Cuba.

References

1. Freud, Sigmund, *The Origins of Psycho-analysis: Letters to Wilhelm Fliess*, edited by Marie Bonaparte, Anna Freud, and Ernst Kris, translated by Eric Musbagher and James Strachey; Basic Books, New York, 1954.
2. Gardner, Martin, *Mathematical Carnival*, Knopf, New York, 1975.

3. ———, ibid., reprinted with further additions, Mathematical Association of America, Washington, 1989.
4. *Chronobiology International* **1** (1984) #2, 103–106.
5. *Australian Journal of Clinical Hypnotherapy and Hypnosis* **6** (1985) #1, 5–13.
6. *Revista de Psihologie* **30** (1984) #2, 166–171.
7. *Journal of Clinical Psychology* **28** (1982) #3, 256–273.
8. *Revista de Psihologie* **28** (1982) #3, 256–273.
9. *Journal of Safety Research* **14** (1983) #4, 167–172.
10. *Psychological Reports* **53** (1983) #2, 612–614.
11. *Przeglad Psychologiczny* **24** (1981) #4, 805–814.
12. *Journal of Sport Behavior* **5** (1982) #3, 132–138.
13. *Psychological Reports* **50** (1982) #2, 396–398.
14. *Journal of Psychology* **110** (1982) #1 39–41.
15. *American Journal of Psychiatry* **128** (1982) #9, 1188–1192.
16. *Aviation, Space, and Environmental Medicine* **51** (1980) #6, 583–590.
17. *Applied Ergonomics* **14** (1983) #3, 215–217.
18. *Perceptual and Motor Skills* **48** (1979) #2, 373–374.
19. *Boletin de Psicologia Cuba* **4** (1981) #1, 13–23.
20. *Boletin de Psicologia Cuba* **6** (1983) #3 80–86.
21. *Revista del Hospital Psiquiatrico de La Habana* **25** (1984) #1 87–93.
22. *Journal of the Society for Psychical Research* **57** (1991) #823, 412–425.

CHAPTER 32

Riding the Wave

The Elliott Wave Theory, a method of predicting the future that is a mixture of science, intuition, and numerology (with considerably less of the first ingredient than the last two) demonstrates, among other things, the immense attractive power of mathematics.

The EWT, discovered by Ralph Nelson Elliott in the 1930s, attempts to explain the behavior of the stock market. Stock prices, Elliott said, move in waves. If prices are going up, there will be an up-wave with five subwaves, three up and two down, as in Figure 1. These will be followed by a down-wave with three subwaves, two down and one up, as in Figure 2. In the up-wave, the three peaks are successively higher and the two troughs, at the end of segments 2 and 4, do not descend as low as the starting point of the previous upward-going segments, 1 and 3. Similarly in the down-wave, the upward segment 2 does not rise as high as the start of the downward segment 1. Nor does the down-wave decline below the point at which the up-wave began.

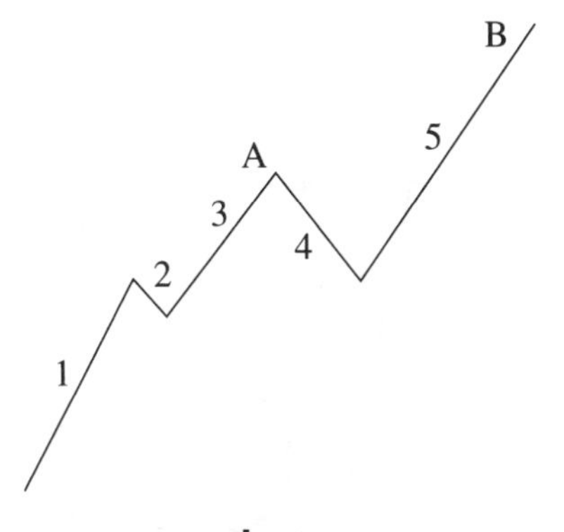

FIGURE 1 An up-wave.

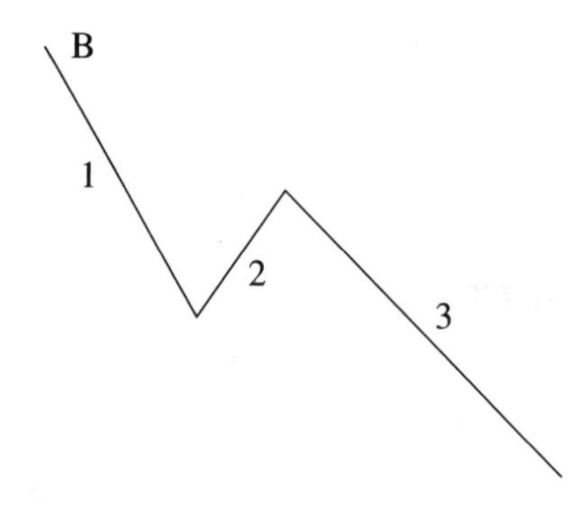

FIGURE 2 The following down-wave.

All of these are qualitative statements—segment 2 in Figure 2 could rise *almost* as high as the start of segment 1, or it could rise hardly at all. Also, its slope could be very large or very small, and the length of time for it to reach its peak is not something the EWT predicts. It gives the rough shape of the curve and none of the details.

If prices are in a downtrend, all is reflected: there will be five waves down and three up (Figure 3).

There are waves within waves. When the trend is up, the up-waves will have five subwaves and the downwaves will have three. When the trend is down, the downwaves will have five subwaves and the up waves will have three. So, we have Figure 4, where the eight segments in Figures 1 and 2 have been joined and subdivided. Segments 1, 3 and 5 in Figure 1 have five subwaves each and segments 2 and 4 three, so the part of Figure 4 from the beginning to the peak has $3 \cdot 5 + 2 \cdot 3 = 21$ segments. The decreasing part, from segments 1, 2, and 3 in Figure 2, has five subwaves in each of the two decreasing segments and three subwaves in the increasing segment, or $5 \cdot 2 + 3 \cdot 1 = 13$ in all. The eight

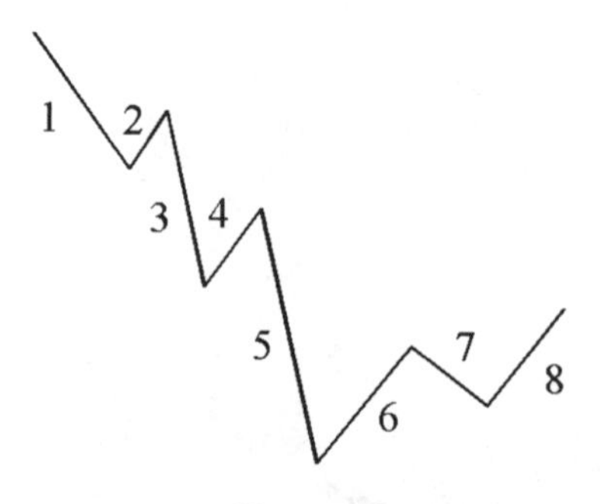

FIGURE 3 A downtrend.

FIGURE 4 A subdivided uptrend.

segments of Figures 1 and 2 thus give rise to $21 + 13 = 34$ subsegments. A further subdivision would give $89 + 55 = 144$ subsubsegments.

If those integers look familiar they should be, since they are elements of the Fibonacci sequence

$$1, 1, 2, 3, 5, 8, 13, 21, 34, 55, 89, 144, 233, 377, \ldots.$$

Since the Fibonacci numbers are very important in the EWT ([1, p. 78] has a picture of Fibonacci's statue in Pisa, taken by one of the authors), here is a quick review of some of their properties. This may be skipped by experts.

The Fibonacci numbers come from a silly problem about the reproduction of rabbits that appeared in the *Liber Abaci*, a book published in 1202 and written by Leonardo of Pisa, commonly referred to today as Fibonacci. (Silly mathematical problems date back at least to the Rhind papyrus, c. 1650 B.C. It is impossible to learn mathematics without them.) The solution of the problem was that the number of pairs of rabbits at the end of n months is f_n, where new terms in the sequence $\{f_n\}$ can be determined from old ones:

$$f_{n+1} = f_n + f_{n-1}, \quad f_1 = f_2 = 1.$$

Besides the recursion relation, there is a formula for f_n:

$$f_n + \frac{\varphi^n - \overline{\varphi}^n}{\sqrt{5}},$$

where

$$\varphi = \frac{1 + \sqrt{5}}{2}, \quad \overline{\varphi} = \frac{1 - \sqrt{5}}{2}.$$

$\varphi = 1.6180339\ldots$ is the famous *golden ratio* about which a good deal of nonsense has been written, both numerological and otherwise (see chapter 29), but which arises naturally. If you ask the question, as the ancient Greeks did, what proportions a rectangle should have so that if a square is cut out of it,

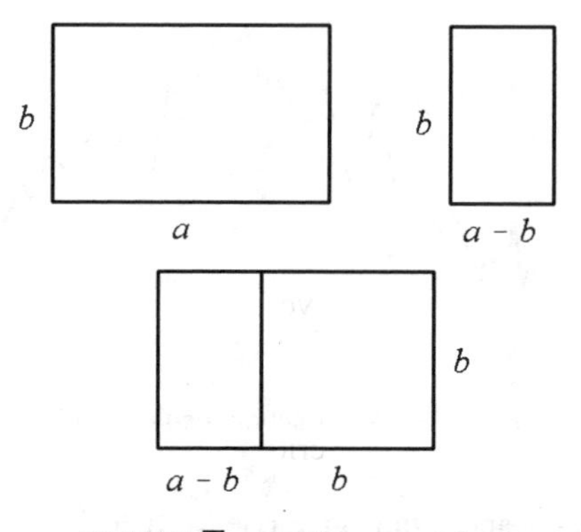

FIGURE 5 Similar rectangles.

the rectangle remaining is similar to the uncut rectangle, the answer is that the ratio of its sides should be φ to 1.

In Figure 5, the original rectangle has dimensions a by b and after the b by b square has been removed, the rectangle that remains is b by $a - b$. So, if the rectangles are similar,

$$\frac{a}{b} = \frac{b}{a-b}.$$

This becomes

$$a^2 - ab = b^2$$

or

$$\left(\frac{a}{b}\right)^2 - \left(\frac{a}{b}\right) = 1,$$

from which it follows that $a/b = \varphi$.

Since $\overline{\varphi} = .6180339\ldots$, $\overline{\varphi}^n \to 0$ as $n \to \infty$, so for n large,

$$f_n = \frac{\varphi^n - \overline{\varphi}^n}{\sqrt{5}} \approx \frac{\varphi^n}{\sqrt{5}}.$$

Thus

$$\frac{f_{n+1}}{f_n} \approx \frac{\varphi^{n+1}/\sqrt{5}}{\varphi^n/\sqrt{5}} = \varphi,$$

so the ratio of successive Fibonacci numbers approaches the golden ratio. The approximation is not bad even for small n:

$$\frac{89}{55} = 1.618181\ldots \quad \text{and} \quad \frac{377}{233} = 1.618025\ldots.$$

Another thing to notice about φ is that

$$\varphi^2 = \frac{1 + 2\sqrt{5} + 5}{4} = \frac{3 + \sqrt{5}}{2} = \varphi + 1.$$

We will return, as do the Elliott wave theorists, to the Fibonacci numbers later. They can be more fascinating even than the stock market.

With the EWT as your guide, you are ready to start making profits, or perhaps losses, betting on stock prices. If you know where you are on the Elliott wave, you know where prices are going to go and, while the theory does not specify exactly either the period or the amplitude of the waves, if you know that the stock market, or your stock (the EWT applies equally well to individuals as to aggregates) is somewhere on segment 2 in Figure 6, the period and amplitude do not matter. Plunge in and wait. You will be a winner. You cannot say precisely when you will reach point A, but you do not need such accurate timing. Just wait until you are on segment 5—it will be clear when that occurs—and then you can sell at a profit at any point on it. Prudent speculators will bail out as soon as the price gets above its value at A because at any time after that segment 5 may terminate and there is no telling exactly when it will. When it does, the descending segment that follows may go down so far and so fast that all gains are wiped out before the imprudent speculator has time to act. Those more daring can hold on longer, trying to get closer to B, but the longer the wait the greater the risk.

The EWT is not all numerology. It is based on an idea of Charles H. Dow, who postulated that

> the market in its primary uptrend was characterized by three upward swings. The first swing he attributed to a rebound from the price over-pessimism of the preceding primary downswing; the second upward swing geared into the improving business and earnings picture; the third and last swing was a price overdiscounting of value. [1, p. 10]

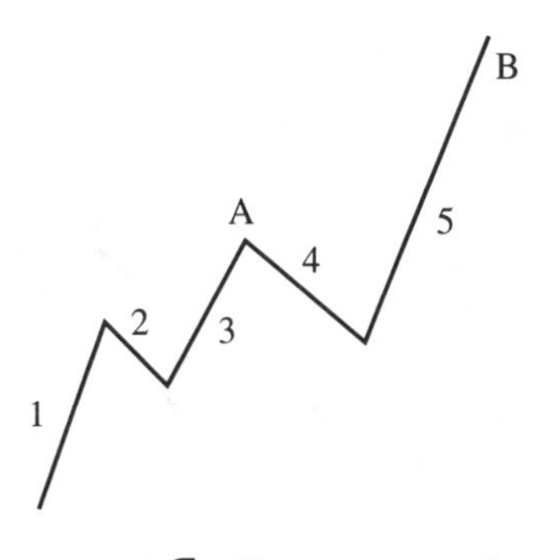

FIGURE 6 The up-wave again.

In Figure 6, segment 1 is the rebound, segment 3 the reaction to improving conditions, and segment 5 the overoptimistic overshooting.

Similarly, in a downtrend we have, successively, reaction to overoptimism, adjustment to lower earnings, and excessive pessimism. In between the three segments, prices drift in the opposite direction, but not so far as to overcome the movement in the primary direction.

The idea has plausibility. Though no statistical test has been able to distinguish the movement of stock prices from a random walk, that does not mean that they are moving at random. In fact, they are not. They are determined by a multitude of individual decisions and we are not smart enough, or lack sufficient information, to discover the cause of their movement. As the digits of π flash by they may seem random to us, but they are anything but; they are completely determined and our inability to say what its 100,000,000th decimal digit is with any more than a one-in-ten chance of being right is a measure of our ineptitude. We do not conclude that the digits of π are random.

Stock market prices are not determined with as much certainty as the digits of π, but it is possible—it is reasonable, even—that *something* should be behind them. Maybe it is the Elliott wave.

Even if it is, the difficulty in applying the EWT is that the charts are not always clear, and can be misinterpreted. For example, suppose we are at point J in Figure 7. Just where are we? ABCD certainly looks like a three-wave down that should be followed by a five-wave up. But maybe DEFGHI *was* the five-wave up. Then we are starting another three-wave down and so should sell short and profit from the coming decline in prices. But, it is also possible that the segments in DEFG were the first three waves of the five-wave up and GHIJ was the fourth wave, the down one, and the wiggles at H and I were irrelevant subwiggles (the curve of prices is self-similar, so we have to decide on our scale of magnification). It that case, prices are due for a big rise and we should bet all we have on an increase. One of the two views will turn out

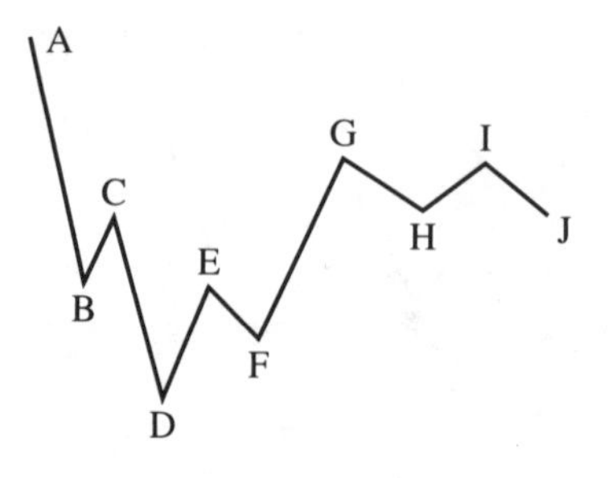

FIGURE 7 Part of a wave.

to be correct, and the EW theorist, on looking at the chart after the increase or decrease has occurred, can point out exactly why what happened in fact had to happen. Whichever way it turns out, the theorist can do that.

EW theorists have thought about this problem:

> Often one will hear several different interpretations of the market's Elliott Wave status, especially when cursory, off-the-cuff studies of the averages are made by latter day experts. However, most uncertainties can be avoided if hourly charts are kept, both on arithmetic and semilogarithmic chart paper, and if care is taken to follow that rules and guidelines as laid down by Elliott. [1, p. 18]

Hourly charts! On two kinds of paper! You would be so busy keeping your charts up to date that you would have no time to place any bets. Of course, this would ensure that you would not lose any money.

The theory is very elaborate. There are impulse waves, diagonal triangles, "truncated fifths (failures)," corrective waves, heads and shoulders, flats, zigzags, double and triple threes, channels, and so on. I have no doubt that EW theorists can, and perhaps have, explained their failures by saying that they mistook a double three for a truncated fifth, or that what seemed to be a flat was really a zigzag.

Even idealized charts can be tough. There follow two, roughly copied from Frost and Prechter [1, p. 166]. In Figure 8 is a neckline—when you have a head, H, and two shoulders, S and S, you naturally have a neckline—from which prices will naturally decline, down to the waist or even to the knees. The start of the decline is shown in the Figure. Figure 9 is a picture of a "false neckline" from which, though there appears to be a head and shoulders, the neckline being false, prices will rise (the start of the rise is indicated). If the fall and rise were not in the charts, could you have told the difference between the two? Evidently EW theorists can, though I was not able to generalize from

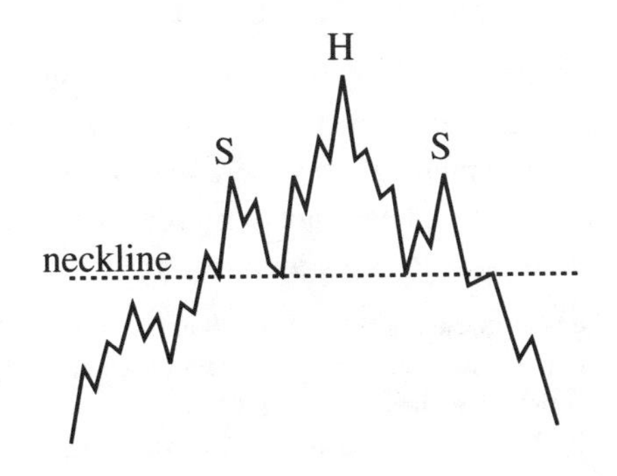

FIGURE 8 A neckline.

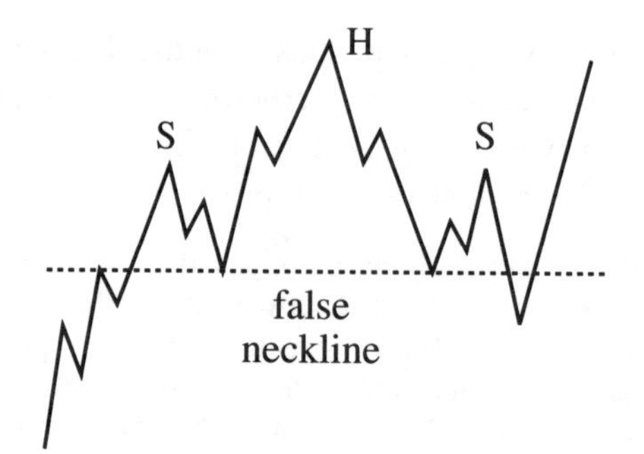

FIGURE 9 A false neckline.

one example the general principle that distinguishes false from true necklines. Even if you had the general principle, applying it to an actual chart would not be easy because of all the subsidiary wiggles, and subsidiaries of subsidiaries.

Nevertheless, EW theorists are not afraid to make what seem to be definite and fearless predictions. For example, here is one, written in 1989:

> Since 1980, *The Elliott Wave Theorist* has made a case that gold has been tracing out a major bear market pattern. The best interpretation of the wave status has been that the first phase of the bear market ended in June 1982 at the termination of an A-B-C decline. From there, gold began an intra-bear market recovery pattern (to be labelled X). The partial recovery pattern, which has retraced a Fibonacci 38.2% of the A-B-C decline

[$\overline{\varphi} = (1 - \sqrt{5})/2$, so $\overline{\varphi}^2 = .381966\ldots$, and that is where "the Fibonacci 38.2%" comes from]

> is either an Ⓐ-Ⓑ-Ⓒ which ended in December 1987 or an Ⓐ-Ⓑ-Ⓒ-Ⓓ-Ⓔ triangle which requires one more rally (though still peaking *below* the December 1987 high of $502.75). From the recovery pattern's end, gold is expected to trace out a *second* A-B-C decline into the early 1990s to complete a "double zigzag" bear market pattern from the 1980 peak at $850/oz. (Any move above $502.75 will negate this analysis.)

This gives a sample of the technical jargon that must be mastered before you can have a full appreciation of the system. An "A" with a circle around it is not the same thing as a plain "A" any more than $f(x)$ is the same as $f'(x)$. It goes without saying that if you have not mastered the technical language, then you cannot expect to be able to apply it correctly.

> To complete the bear market, gold should fall at minimum to below $200/oz. This projection appears outrageously low relative to current generally held expectations, particularly since such expectations are for *higher* prices, not lower ones at all. Because prices usually fall close to the *low* of the previous fourth wave (in this case, $103.50), the strongest possibility for an ultimate

bottom is near the *lower end* of the $103.50–$197.50 range. The aftermath could be a dramatic re-inflation, or even hyperinflation, probably mid-decade. [1, p. 240]

In spite of mentioning specific prices, the prediction is not as fearless as it may seem. What it is really saying is that the price of gold will, at some time in the future, either be more than $502.75 an ounce or less than $200 an ounce. This is a safe prediction. In fact, it cannot be wrong. If you assert that the price of *anything* will, at some time, break outside *any* band, you will be correct, with probability 1.

The Fibonacci numbers that arise when you subdivide waves into smaller waves impressed Messrs. F. and P. very much. The book has a whole chapter on the mathematical background of the wave principle. There is a summary of Fibonacci's life and work, a description of the rabbit problem, and the introduction of φ as the limit of the ratio of successive Fibonacci numbers. The authors are evidently impressed with the properties of some numbers associated with the Fibonacci sequence and give a list of them [1, p. 86]:

1) $2.618 - 1.618 = 1.$
2) $1.618 - .618 = 1.$
3) $1 - .618 = .382.$
4) $2.618 \times .382 = 1.$
5) $2.618 \times .618 = 1.618.$
6) $1.618 \times .618 = 1.$
7) $.618 \times .618 = .382.$
8) $1.618 \times 1.618 = 2.618.$

Since the numbers are φ (1.618), φ^2 (2.168), φ^{-1} (.618), and φ^{-2} (.382), the properties are not a surprise to those who know that $\varphi^2 - \varphi = 1$ and who remember how to multiply powers since they are

1) $\varphi^2 - \varphi = 1$
2) $\varphi - \varphi^{-1} = 1$
3) $1 - \varphi^{-1} = \varphi^{-2}$
4) $\varphi^2 \cdot \varphi^{-2} = 1$
5) $\varphi^2 \cdot \varphi^{-1} = \varphi$
6) $\varphi^2 \cdot \varphi^{-2} = 1$
7) $\varphi^{-1} \cdot \varphi^{-1} = \varphi^{-2}$
8) $\varphi \cdot \varphi = \varphi^2.$

Messrs. F. and P. then give some properties of the Fibonacci sequence. They are included not because they have anything to do with Elliott waves, because they do not. They are there because numbers have power, in this case the power to engage and fascinate to such a degree that the authors cannot resist the urge to share their wonderfulness. One property is

> 4) The sum of all Fibonacci numbers in the sequence up to any point, plus 1, equals the Fibonacci number two steps ahead of the last one added.
>
> [1, p. 78]

This is more compactly expressed by

$$f_1 + f_2 + \cdots + f_n + 1 = f_{n+2},$$

but however it is expressed, it is not going to be any help in beating the stock market. Nor is

> 6) The square of any Fibonacci number minus the square of the second number below it in the sequence is always a Fibonacci number.

So it is:

$$3^2 - 1^2 = 8, \quad 5^2 - 2^2 = 21, \quad 8^2 - 3^2 = 55, \quad 13^2 - 5^2 = 144, \ldots$$

or

$$f_4^2 - f_2^2 = f_6, \quad f_5^2 - f_3^2 = f_8, \quad f_6^2 - f_4^2 = f_{10}, \quad f_7^2 - f_5^2 = f_{12}, \ldots$$

and in general

$$f_n^2 - f_{n-2}^2 = f_{2n-2}.$$

Messrs. F. and P. were so taken with Fibonacci numbers that they did research, and made a discovery. This is what numbers can do, and have done, throughout history: they are so fascinating, and have so much power, that they draw people in to investigate them, for their own sake. Here is the discovery:

> 8) One mind stretching phenomenon, which to our knowledge has not previously been mentioned, is that the ratios between Fibonacci numbers yield numbers which very nearly are thousandths of other Fibonacci numbers, the difference being a thousandth of a third Fibonacci number, all in sequence. Thus, in ascending direction, identical Fibonacci numbers are related by 1.00 or .987 plus .013, adjacent Fibonacci numbers are related by 1.618, or 1.597 plus .021, alternate Fibonacci numbers are related by 2.618, or 2.464 plus .034, and so on.... On all counts, we truly have a creation of "like from like," of "reproduction in an endless series," revealing the properties of "the most binding of all mathematical relations," as its admirers have characterized it.
>
> [1, pp. 87–88]

What Messrs. F. and P. noticed is that, to three decimal places,

$$\varphi^0 = 1 = .987 + .013 = \frac{f_{16} + f_7}{1000}$$

$$\varphi^1 = 1.618 = 1.597 + .021 = \frac{f_{17} + f_8}{1000}$$

and

$$\varphi^2 = 2.618 = 2.464 + .034 = \frac{f_{18} + f_9}{1000}.$$

The general assertion is that

$$\varphi^n = \frac{f_{n+16} + f_{n+7}}{1000}, \quad n = 0, 1, 2, \ldots,$$

at least to three places of decimals. That would be a notable discovery if it were correct, so it is too bad that it is not. The first exception does not occur until $n = 9$ where, to three places,

$$\varphi^9 = 76.013 \quad \text{and} \quad \frac{f_{25} + f_{16}}{1000} = 76.012.$$

The reason for the near equalities is that $f_{16} + f_7 = 987 + 13 = 1000$. Since

$$f_n \approx \frac{\varphi^n}{\sqrt{5}},$$

we have

$$\begin{aligned} f_{n+16} + f_{n+7} &\approx \frac{\varphi^{n+16} + \varphi^{n+7}}{\sqrt{5}} = \frac{\varphi^n}{\sqrt{5}}(\varphi^{16} + \varphi^7) \\ &\approx f_n \frac{f_{16} + f_7}{\sqrt{5}} = 1000 \frac{f_n}{\sqrt{5}} \approx 1000\varphi^n \end{aligned}$$

which is the discovery.

Numbers have such power that, when it comes to a choice between numbers and reality, the numbers win. Fibonacci numbers rule.

> Modern science is rapidly discovering that there is indeed a basic proportional principle of nature. By the way, you are holding this book with two of your *five* appendages, which have *three* jointed parts, *five* digits at the end, and *three* jointed sections to each digit. [1, p. 89]

What is more, the book cost me *thirty-four* dollars, plus sales tax at *five* percent. This may be the *fifty-fifth* illustration (or even the *eighty-ninth*) of the operation of the Law of Small Numbers in this book.

Messrs. F. and P. were also enamored of the golden ratio:

> The Golden Section occurs throughout nature. In fact, the human body is a tapestry of Golden Sections (see Figure 70) in everything from outer dimensions to facial arrangement. "Plato, in his *Timaeus*," says Peter Tompkins, "went so far as to consider *phi*, and the resulting Golden Section proportion, the most binding of all mathematical relations, and considers it the key to the physics of the cosmos." [1, pp. 89–90]

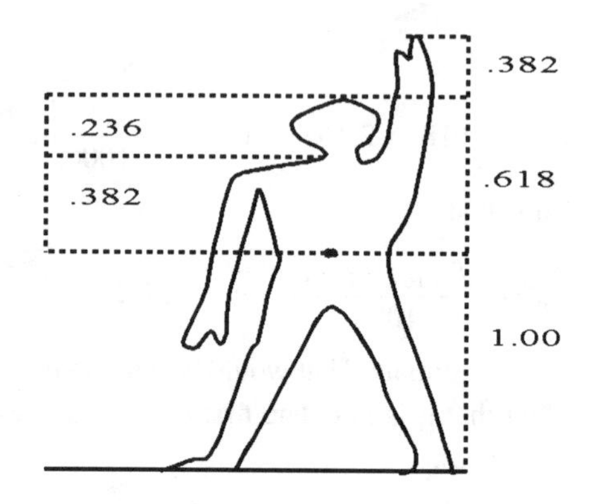

FIGURE 10 Adaptation of Figure 70 from [1].

Figure 10 is a rough copy of the authors' Figure 70. The black circle is the navel. (Peter Tompkins is the great pyramid man whose findings were considered in chapter 27. Calling φ a "mathematical relation" is strange to mathematical ears and how a mathematical relation can possibly "bind" is a mystery.)

Mysticism creeps in, a strange thing to find in a book on how to make money in the stock market:

> The Pythagorean brotherhood chose the five pointed star as their symbol, as every segment of the figure is in golden ratio to the next smaller segment. In the sixteenth century, Johannes Kepler, in writing about the Golden, or "Divine Section," said that it described virtually all of creation, and specifically symbolized God's creation of "like from like." Man's most reliable outer body relationship is the division made at the navel. The statistical average is exactly .618. The ratio holds true separately for men, and separately for women, a divine proportion indeed and a fine symbol of the creation of "like from like." Is all of mankind's progress a creation of "like from like?"...
>
> Works of art have been greatly enhanced with knowledge of the Golden Rectangle. Fascination with its value and use was particularly strong in ancient Egypt and Greece and during the Renaissance, all high points of civilization. [1, p. 90, 92]

There is no evidence that the ancient Egyptians knew a thing about φ, and there is even the possibility that their navels were situated differently from those of modern writers on the golden ratio. But that is not allowed to get in the way of a good story:

> The ancient Egyptians consciously enshrined forever the Golden Ratio in the Great Pyramid of Gizeh by giving its faces a slope height equal to 1.618 times half the base, so that the vertical height of the pyramid is at the same time the square root of 1.618 times half its base. According to Peter Tompkins...
> [1, p. 100]

(See chapter 23 for a more reasonable explanation of the slope of the pyramid.)

> Tompkins explains. "The Pharaonic Egyptians," says Schwaller de Lubicz, "considered *phi* not as a number, but as a symbol of the creative function, or of reproduction in an endless series. To them it represented 'the fire of life, the male action of sperm, the *logos* [referenced in] the gospel of St. John.' " [1, pp. 100–101]

Since the ancient Egyptians never wrote a word about φ—they couldn't, because their crude mathematics was not up to it—it is hard to know where Schwaller came by his information. In any event, Messrs. F. and P. conclude that φ is built into the stock market:

> If natural law permeates the universe, should it not permeate the world of man as well? If the order of the universe and of man's body and brain reflect the form of *phi*, might not man's actions be similarly tied, by his nature? If man's progress is based upon production and reproduction "in an endless series," is it not possible that such progress has the spiralling form of *phi*, and that this form is discernible in the movement of the stock market? The stock market, in our opinion, can be understood if it is taken for what it *is* rather than what it *seems* it should be on the surface, that is, not a formless mass reacting to current news events, but a record of the formal structure of the progress of man. [1, p. 101]

Messrs. F. and P. have come a long way from the observation that advances in the stock market might be divided into three parts, the correction, the realization, and the overextension. That observation, along with the assertion that the cycles contain within them smaller cycles which exhibit the same form, lead to the Fibonacci numbers. Fibonacci numbers are wonderful and led, as numbers have the power to do, to the addling of the authors' minds. Once minds have been so addled, they see numbers everywhere. For example, [1, p. 121]:

> In *Nature's Law*, Elliott gave the following examples of Fibonacci time spans between important turning points in the market:

1921 to 1929	8 years
July 1921 to November 1928	89 months
September 1929 to July 1932	34 months
July 1932 to July 1933	13 months
July 1933 to July 1934	13 months
July 1934 to March 1937	34 months
July 1932 to March 1937	55 months
March 1937 to March 1938	13 months
March 1937 to April 1942	5 years
1929 to 1942	13 years

Those turning points have probably been forgotten, but the memory of the sharp decline in the stock market in 1987 may still be fresh. Fibonacci numbers pointed to that as well:

> *Elliott Wave Principle* pinpointed 1987 as having the most precise lineup of coincidentally terminating Fibonacci time lengths within the Supercycle from 1932:
>
> $$\begin{aligned} 1932 \text{ (major low)} + \underline{55} \text{ years} &= 1987 \\ 1953 \text{ (minor low)} + \underline{34} \text{ years} &= 1987 \\ 1966 \text{ (major top)} + \underline{21} \text{ years} &= 1987 \\ 1974 \text{ (major low)} + \underline{13} \text{ years} &= 1987. \end{aligned}$$
>
> At the time, all we could be sure of was that 1987 would provide a major turning point (we suspected a bottom). In 1982, when a major low appeared to have occurred, the string of Fibonacci projections expanded, since 1982 + 5 years pointed to 1987 as well. (Shortly thereafter, the deepest correction within the bull market ended in 1984, which was *3* years from 1987.) [1, pp. 221–222]

There is no reason given for why these Fibonacci numbers occur, other than the appeal to φ as a binding mathematical relationship.

Messrs. F. and P. have some odd ideas about mathematics as well [1, p. 203]:

> The calculus and the imaginary are both concerned with the angle of rotation. Integration corresponds to a positive rotation and differentiation to a negative rotation with respect to the reference. Differentiation represents the rotations of a tangent to a point moving along a curve. Integration represents a reconstruction of the line from the rotations of the tangent.
>
> $$\begin{aligned} &\int^{-1} (\cos) = -\sin \qquad && i^{-1}(1) = -i \\ &\int (\sin) = -\cos && i^{-1}(-i) = -1 \\ &\int (-\cos) = \sin && i^{-1}(-1) = i \\ &\int (\sin) = \cos && i^{-1}(i) = 1 \end{aligned}$$

I have copied those equations exactly as they appear, though there is a conflict between the second and the fourth. Perhaps one of the integral signs has a missing exponent. Since those equations come just after

$$\begin{aligned} bi &= be^{\frac{\pi}{2}i} \\ &= be\cos\frac{\pi}{2} + bi\sin\frac{\pi}{2} \\ &= 0 + bi \end{aligned}$$

(equations also copied exactly as they appear), one is not impressed with the power of the authors' mathematics.

In common with many other people who do funny things that are related to mathematics, Messrs. F. and P. invoke the name of Kurt Gödel (with a new spelling) to try to justify, partly, what they have done.

> Goedal demonstrated in his Proof that the resources of the human mind cannot be formalized completely and that new principles may always be found by discovery and by pragmatic methods. [1, p. 204]

Whether Gödel demonstrated that or not, it does not imply that the EWT is correct.

However shaky the EW people may be on Fibonacci numbers, navels, pyramids, integrals of trigonometric functions, and the spelling of Gödel, they are firm on price. I wanted to examine some of the works of Ralph N. Elliott himself, but on looking them up in *Books in Print* I found

> Elliott, R. N., *Nature's Law and the Secret of the Universe*, 1980, 327.50.
>
> ———, *Nature's Law*, 105p, 1975, 377.75.
>
> Elliott, R. N. and Fleming, Spencer, *The Fibonacci Rhythm Theory as It Applies to Life, History, and the Stock Market*, 2 vols, 1985, 427.00.
>
> Fibonacci, Leonardo and Elliott, Ralph N., *The Fibonacci Rhythm Theory*, ed. by Clifford Griffith, 157p, 1980, 427.00.

Look at those prices! I would not pay them, even to get a new work by Fibonacci, his first since 1250.

I have an advertisement for the EW Precision Ratio Calipers ("[A] perfect gift for clients and friends") which reads:

> **Makes Fibonacci and Gann Analysis easy**.... With the Precision Ratio Compass you'll save time, obtain greater accuracy, and project *price and time* targets with ease. You can quickly and efficiently mark your charts with retracements, multiples, third and fourth generation Fibonacci ratios *and all other ratios as well*.... The reference manual includes *never-before published uses of the Fibonacci ratio*...

The PRC is a metal device, seven inches long, consisting of two slotted bars joined with a thumbscrew with logarithmic scales down their sides, one marked "circles" and the other "lines," and sharp points at both ends. If the EWT ever caught on nationwide in a big way, I have no doubt that cheap versions of the PRC could be found at Wal-Mart for $5.95 or less. But to get one from Elliott Wave International, P. O. Box 1618 (Note the box number—φ! Is that not a nice touch?), Gainesville, Georgia, 30503, you will have to come up with $249, plus $3 for shipping and handling.

The lesson is that there is more than one way to get rich in the stock market.

Reference

1. Prechter, Robert Rougelot and Alfred John Frost, *The Elliott Wave Principle*, first edition 1978, sixth edition (expanded) 1990, New Classics Library, Inc., Gainesville, Georgia.

CHAPTER 33
Conclusion

There is no conclusion. There can be no conclusion as long as human folly persists, which will be for the lifetime of the race. We are blessed and cursed with minds that possess selfconsciousness and intelligence, so some of us can look at the world and, in spite of its manifold wonders, be dissatisfied with it. Our lives proceed from birth to death, but for some of us their prosaic joys are insufficient. Some of us want more. We want marvels and mysteries, we want to find secret knowledge, we want to live in the end times, we want *something* more than the dreary ordinary existence that is our lot. So, some of us become numerologists and others of us their followers.

Well, what of it? If some people want to have harmless delusions, why not let them? Why should you, and I, go to the trouble of pointing out the lack of sense in nonsense and of ridiculing the ridiculous?

Because we need to. Because we must. Now is the golden age of humanity, you know. In spite of the enormous amount of pain in the world, the innumerable unknown tragedies, the injustice, the squalor, the horrors, this is, for more people than at any time in history, the best time to be alive. I am glad to be alive now, and I want my children, and their children, to live in a world as marvelous.

To preserve what we have takes effort, constant effort. The dark forces of chaos and unreason surround us. They are always pushing, relentlessly probing, never ceasing in their efforts to tear down the fragile edifice of knowledge, reason, and civilization that we have created. If we do not resist them, they will advance, and advance, and the light will dim and the golden age will end.

Every convert to numerology advances the darkness. Each new pyramid-measurer is a defeat. Every biorhythm chart is a blow. Human ignorance, credulity, and superstition will not go away, but we must do what we can to keep them in check. Otherwise they could overwhelm us.

Index